Arms Control and the Rule of Law: A Framework for Peace and Security in Outer Space

Proceedings of the Fifteenth Annual Ottawa NACD Verification Symposium

Edited by: J. Marshall Beier and Steven Mataija

CENTRE FOR INTERNATIONAL
AND SECURITY STUDIES
York University
Toronto, Canada
1998

THE YORK CENTRE FOR INTERNATIONAL AND SECURITY STUDIES supports and encourages innovative theoretical and empirical research in international relations and security studies. The Centre fosters scholarly research of interest to academic and policy communities in Canada and abroad. The Centre also coordinates and facilitates a variety of scholarly activities including conferences, workshops, lectures, seminars, and research publications.

Research at the Centre falls within four streams: (1) Strategic Studies; (2) Political Economy of Security; (3) Regional Conflict; and (4) Non-Traditional Analyses of Conflict and Security.

For further information please contact:

Centre for International and Security Studies
3rd Floor York Lanes
York University
4700 Keele Street
Toronto, Ontario
Canada M3J 1P3

Telephone:	(416) 736–5156
Fax:	(416) 736–5752
E-mail:	yciss@yorku.ca
Internet:	http://www.yorku.ca/research/ciss

Canadian Cataloguing in Publication Data

Ottawa NACD Verification Symposium (15th : 1998 : Montebello, Quebec)
Arms control and the rule of law : a framework for peace and security in outer space

Proceedings of a symposium held in Montebello, Quebec, Mar. 11–14, 1998.
Includes bibliographical references
ISBN 0–920231–26–8

1. Arms control – Congresses. 2. Arms control – Verification – Congresses.
3. Astronautics, Military – Congresses. I. Beier, J. Marshall, 1966– .
II. Mataija, Steven, 1968– . III. York Centre for International and Security Studies. IV. Title.

UA12.5.077 1998 341.7′3 C98–931860–5

Dedicated to the memory of
Sidney N. Graybeal,
friend and colleague

Table of Contents

List of Contributors vii

Preface ix

Editorial Foreword xi

Part I. Space Applications: The Existing Regime and Prospects for the Future

Chapter One — Outer Space and Multilateral Security: Current Trends and Possibilities
Freleigh J.F. Osborne 3

Chapter Two — The Relationship of the ABM Treaty to Strategic Stability and Outer Space
Patricia Bliss McFate and Sidney N. Graybeal 15

Chapter Three — Outer Space: The Conference on Disarmament Dimension
Donald Sinclair 29

Chapter Four — Banning the Stationing of Weapons in Space Through Arms Control: A Major Step in the Promotion of Strategic Stability in the 21st Century
F.R. (Ron) Cleminson 35

Chapter Five — Recent Developments in Space Law
M. Lucy Stojak 45

Chapter Six — A Variant of a Mandate for an Ad Hoc Committee on Outer Space within the Conference on Disarmament: A Convention for the Non-Weaponization of Outer Space
Phillip J. Baines 65

Chapter Seven — Confidence Building and Outer Space
James Macintosh 85

Chapter Eight — A Total Ballistic Missile Ban Treaty
Phillip J. Baines 103

Chapter Nine — The Protection of Satellites: Problems for Arms Control
George Lindsey 113

Chapter Ten — The Ballistic Missile Defence and Anti-Satellite Relationship
Phillip J. Baines 121

Chapter Eleven — The Role of Non-Governmental Organizations in Maintaining Space as a Weapons-Free Environment
Peter Stibrany 131

Part II. Extending the Agenda

Chapter Twelve — A View on the Application of Overhead Imagery to Strengthen the IAEA Safeguards Regime
Hui Zhang 145

Chapter Thirteen — Northeast Asian Insecurity: New Sources of Friction Between States and Implications for Arms Control and Disarmament
Hong-Kuan Jiang 163

Chapter Fourteen — Symposium Summary
George Lindsey 189

Part III. Working Group Reports

Chapter Fifteen — Working Group One Rapporteur's Notes: Legal and Policy Concerns
Peter Gizewski 199

Chapter Sixteen — Working Group Two Rapporteur's Notes: Arms Control in Outer Space — The Multilateral Dimension
Jean-Philippe Tachdjian 203

Chapter Seventeen — Working Group Three Rapporteur's Notes: Outer Space and Strategic Stability — The ABM Treaty and Ballistic Missile Defence
Steve James 207

Chapter Eighteen — Working Group Four Rapporteur's Notes: Building a Constituency to Inform the Debate
Peter Stibrany 211

List of Abbreviations and Acronyms 215

List of Contributors

PHILLIP J. BAINES, Verification Research Officer, Non-Proliferation, Arms Control and Disarmament Division, Department of Foreign Affairs and International Trade, Ottawa, Canada

J. MARSHALL BEIER, Department of Political Science, Centre for International and Security Studies, York University, Toronto, Canada

F.R. (RON) CLEMINSON, Senior Advisor on Verification, Non-Proliferation, Arms Control and Disarmament Division, Department of Foreign Affairs and International Trade, Ottawa, Canada

PETER GIZEWSKI, NACD Research Associate, Centre for International and Security Studies, York University, Toronto, Canada

SIDNEY N. GRAYBEAL, Science Applications International Corporation (SAIC), Sante Fe, United States

STEVE JAMES, Space Analyst and Instructor, Canadian Forces School of Aerospace Studies, Winnipeg, Canada

HONG-KUAN JIANG, DFAIT Research Fellow, Department of Political Science, University of Victoria, Victoria, Canada

GEORGE LINDSEY, Vice President (Research Programmes), Canadian Institute of Strategic Studies, Ottawa, Canada

JAMES MACINTOSH, Canadian Security Research, Elmira, Canada

STEVEN MATAIJA, Conference and Information Coordinator, Centre for International and Security Studies, York University, Toronto, Canada

PATRICIA BLISS McFATE, Senior Scientist and Program Director, Center for Global Security and Cooperation, Sante Fe, United States

FRELEIGH J.F. OSBORNE, F.J.F. Osborne & Associates, Beaconsfield, Canada

DONALD SINCLAIR, Permanent Mission of Canada to the Office of the United Nations at Geneva, to the Conference on Disarmament and to the World Trade Organization, Geneva, Switzerland

PETER STIBRANY, Iconitec Multimedia Inc., Vaudreuil-Dorion, Canada

M. LUCY STOJAK, Director, Summer Session Program (SSP), International Space University, Illkirch-Graffenstaden, France

JEAN-PHILIPPE TACHDJIAN, Legal Operations Division, UN Legal Affairs, Department of Foreign Affairs and International Trade, Ottawa, Canada

HUI ZHANG, Post-Doctoral Research Fellow, Center for Energy and Environmental Studies, Princeton University, Princeton, United States

Preface

This, the Fifteenth Annual Ottawa NACD Verification Symposium, can be seen as something of a benchmark in the series. It is representative of the quality of research which has been conducted over the years in the areas of monitoring, verification, confidence building, and transparency as it relates to Canada's non-proliferation, arms control, and disarmament (NACD) agenda. Since its inception in 1983, the Department of Foreign Affairs' Verification Research Program (VERPROG) has focussed on both concept and practice in these areas. Monitoring and verification are recognized as essential ingredients in the successful completion and implementation of NACD treaties and agreements.

VERPROG has fostered a sort of "made-in-Canada" approach to the processes of monitoring and verification, as well as to confidence building and the enhancement of transparency. It has shared its research results with allies, friends, and others who have expressed interest in a multilateral approach to the NACD process. These symposia have served as something of a "sounding board" to review the results of research at the end of each fiscal year. It has also constituted a means of exchange with NACD experts from other countries. Canadian verification research has benefited greatly from the resultant opportunities to maintain a dialogue with international experts at the tops of their respective professions.

Clearly, as expertise and knowledge has accumulated in these functional areas, the agendas of successive symposia have been broadened to encompass the NACD process in its wider policy and political dimensions, but always with a "process-oriented" approach. Participation has expanded from a fairly close-knit, NATO-centric grouping of the early-1980s to one which now includes colleagues from the Russian Federation, the Peoples' Republic of China, the Republic of Korea, and others 'from a distance.' The common thread more recently has been a shared and abiding interest in multilateral approaches to issues, problems, and solutions in the NACD context.

Over the years, while the agendas have changed to meet new challenges, the symposium coordinators — initially from Carleton University in Ottawa and later from York University in Toronto — have hosted more than 300 experts from home and abroad, some of them attending more than one symposium. Here, two individuals deserve special mention. First, Ambassador Jonathan Dean, whose attendance at the first symposium was instrumental in establishing its credibility. His continued participation, in a variety of capacities, has served as a strong endorsement of its continued validity. Ambassador Dean is a veritable icon in the field of multilateral arms control. Second is Steven Mataija, who has quickly and efficiently 'stage-managed' these symposia for several years. While content is the basis upon which success of a symposium is measured, effective organization, from the initial development of the agenda to final publication of the proceedings, is how success is assured.

The focus and process of these symposia is likely to evolve together with the multilateral NACD agenda in the years ahead. The spirit and integrity of each and the level of expertise of the participants are the constants which will ensure future success. From my own perspective, the proceeds from each of these symposia represent the best distillation of thought and expertise on issues of the time and, often, on those identified for the future.

The Fifteenth Annual Symposium, entitled "Arms Control and the Rule of Law: A Framework for Peace and Security in Outer Space," is no exception. It identifies outer space as a present concern, but one which is also likely to be recognized as the single most important NACD issue in the first decade of 21st Century. Multilateralists have not only a right but a responsibility to be intimately concerned with outer space as an NACD issue of overriding international importance. These proceedings serve as a credible introduction to this issue area. It is to be hoped that the lessons learned in other environments can be applied to a multilateral approach to the outer space issue and that, from an arms control perspective, outer space can be kept weapons free.

Finally, I would like to express my gratitude to all who have participated in past symposia and to express my best wishes to those who attend in the future.

F.R. Cleminson

Editorial Foreword

The chapters included in this volume are based on papers commissioned for and presented at the Fifteenth Annual Ottawa NACD Verification Symposium, sponsored by the Non-Proliferation, Arms Control and Disarmament Division of the Department of Foreign Affairs and International Trade (Canada) and organized by the York Centre for International and Security Studies. The Symposium was held in Montebello, Québec, 11–14 March 1998. A group of 40 experts from government, the military, industry, and academe attended the conference. The volume generally reflects the sequence of presentations made during the Symposium.

We would like to dedicate this volume to the memory of Sidney N. Graybeal. Ambassador Graybeal passed suddenly shortly after attending this year's Symposium. A former arms control negotiator, Sidney also worked for the US Central Intelligence Agency as well as the State and Defense Departments before retiring as director of the CIA's Strategic Research Office in 1979. In a distinguished career which spanned six decades, Sidney's achievements included service on the historic SALT-I negotiating team, the Defense Department's Strategic Weapons Panel, the staff of United Nations Ambassador Arthur Goldberg for the negotiation of treaties dealing with outer space, and the 1968 delegation to the Eighteen Nation Disarmament Conference. He was a senior intelligence adviser at the White House during the Cuban Missile Crisis. Sidney was also the first US commissioner of the Standing Consultative Commission on implementation of the Anti-Ballistic Missile (ABM) Treaty. Sidney's accomplishments were recognized with the bestowment of the Presidential Award for Distinguished Service, the highest honour that can be conferred upon an American civil servant. Following his years of government service, he became a senior vice president with System Planning Corp. and in 1989 began working for Science Applications International Corp., becoming chief scientist. Ambassador Graybeal was a frequent and valued participant and contributor to the annual Ottawa NACD verification symposia. His support of and enthusiasm for the Verification Research Program at the Department of Foreign Affairs and the related activities of the Centre for International and Security Studies at York University will be sadly missed.

This year's Symposium was designed to explore the concept of maintaining and reinforcing peace and security in outer space by taking advantage of the potential strengths which verifiable arms control agreements and the evolution of international law can offer. Chapter One speaks to this thematic in a broad sense. In this chapter, Freleigh Osborne conducts an extensive survey of current trends and future possibilities for multilateral initiatives for security in outer space.

In Chapter Two, Patricia Bliss McFate and Sidney Graybeal discuss the relationship of the ABM Treaty to strategic stability and outer space. They provide us with an overview of the ABM Treaty, including its negotiation background and its implementation and operation, and examine the future of the Treaty. Finally, they identify certain likely events or consequences associated with the continued existence or demise of the Treaty.

Donald Sinclair addresses the question of outer space from the Conference on Disarmament (CD) perspective in Chapter Three. Specifically, he examines what the CD has done in the past with respect to outer space as well as the prospects for work on this issue in the CD this year, and debates whether an Ad Hoc Committee on outer space should be established. Sinclair also makes some suggestions as to Canada's role in these efforts.

In Chapter Four, F.R. (Ron) Cleminson evaluates the adequacy of the existing space law regime to prohibit military uses of or activities in outer space and concludes that the goal of ensuring that space remains non-weaponized is unfulfilled. He notes that there are many loopholes and lacunae in the existing legal framework and many treaty provisions open to a variety of interpretations, and recommends that this is the opportune time for the international community, in its deliberative and negotiating bodies (especially the Conference on Disarmament and COPUOS), to examine the outer space issue in all its aspects and to take appropriate action.

M. Lucy Stojak examines recent developments in space law in Chapter Five, focussing on Global Positioning Systems (GPS), remote sensing, the International Space Station, and the Missile Technology Control Regime (MTCR) guidlines. She also provides us with a summary of various proposals made to amend the Outer Space Treaty, and assesses the likelihood of such amendments actually being made. She concludes with a series of modest recommendations to reactivate interest in the CD on the topic of the prevention of an arms race in outer space.

In Chapter Six, Phillip J. Baines presents the case for a convention for the non-weaponization of outer space. He concludes that such a convention could be equitable, verifiable, and in the national security interests of all nations.

James Macintosh, in Chapter Seven, provides readers with a brief introduction to confidence building and offers several observations about the potential use of confidence building in the context of outer space. His intent is to provide those interested in international law and arms control in outer space with some food for thought relating to the use of the confidence building approach.

In Chapter Eight, Phillip J. Baines presents the case for a total ballistic missile ban treaty. He indicates that by utilizing the resources of improved space surveillance sensors tracking all aerospace objects in flight above the Earth and numerous Earth remote-sensing satellites viewing all terrestrial activity below, together with the demonstrated effectiveness of on-site inspections, it is possible to verify a complete ban on the development, testing, deployment, and use of ballistic missiles and ballistic missile defence systems. What is required by the international community, Baines notes, is the political will to suffocate the technology that fuels the nuclear arms race.

The next two chapters examine the relationship between the anti-satellite mission and the ballistic missile defence mission. Both George Lindsey and Phillip J. Baines conclude that a meaningful ban on the deployment of anti-satellite systems is not possible in the presence of deployment of either theatre or strategic ballistic missile defence systems.

Peter Stibrany makes some propositions regarding the potential benefits of NGO involvement in multilateral space arms control processes in Chapter Eleven. He considers whether NGOs have a role in creating visibility and momentum which could initiate a space arms control process and sustain it to a successful conclusion and whether they have a role in monitoring a multilateral space arms control agreement, should one be concluded.

Part II of the proceedings includes papers from the portion of the Symposium entitled "Extending the Agenda," which provides a forum for considering new ideas and perspectives on arms control and verification issues. In Chapter Twelve, Hui Zhang examines the possibilities for applying present and projected overhead imagery — particularly commercial observation satellites — to strengthen the IAEA safeguards regime. He discusses the synergistic aspects of the use of overhead imagery systems with existing methods of monitoring and verification under the IAEA safeguards programme, and looks at the utility of applying such systems to the monitoring and verification of a possible fissile material cutoff treaty. In addition, some suggestions are made regarding the appropriate organization and management of such a system under the IAEA. In Chapter Thirteen, Hong-Kuan Jiang examines the emerging trends of non-traditional security challenges (NTSCs) in the Sea of Japan and the Yellow Sea basin area, and discusses the implications for potential application of arms control and verification technology. Finally, in Chapter Fourteen, George Lindsey provides us with a summary of some of the key themes which recurred throughout the Symposium.

Part III includes rapporteur's notes from the four working groups conducted at the Symposium. Participants were divided into working groups and provided with questions which were simple and generic, the objective being to foster discussion. Participants were encouraged to address other related issues and the working groups were asked to work toward identifying specific conclusions or recommendations, in addition to general observations. The results of these deliberations can be found in Chapters Fifteen through Eighteen.

We are indebted to all conference participants for their commitment and involvement, and in particular, to John S. MacDonald and Necil Nedimoglu, each of whom gave a dinner address. Special appreciation goes to those authors who have contributed papers for this volume, and to the Non-Proliferation, Arms Control and Disarmament Division, particularly Alan Crawford, Jill Sinclair, Gordon Vachon, and F.R. (Ron)

Cleminson. We would like to extend to Ron both our congratulations on his upcoming retirement and our gratitude for his unrelenting energy, vision, and initiative in establishing Canada as a leader in the field of verification of non-proliferation, arms control, and disarmament agreements. It is our hope that Ron will continue to participate and lend his valued expertise to future symposia.

Finally, we would like to thank our colleagues at the Centre for International and Security Studies for their assistance and support, including Joan Broussard, Heather Chestnutt, David Dewitt, and Wendy Kubasik. It is hoped that these papers will stimulate further interest, debate, and analysis of the issues examined herein. Of course, the editors are responsible for any errors of fact and interpretation.

J. Marshall Beier and Steven Mataija
York University, Toronto
July 1998

Part I
Space Applications: The Existing Regime and Prospects for the Future

Chapter One

Outer Space and Multilateral Security: Current Trends and Possibilities

Freleigh J.F. Osborne*

Introduction[1]

All nations border on space and therefore have legitimate interests and associated, or related, concerns about space. The relevant activities of the technologically advanced nations centre on the ability of space to provide infrastructure capabilities relating to their military, commercial, and social requirements. However, in addressing these requirements many nations, despite their technology, are or have been dependent upon the political, military, and/or commercial policies of other nations. The corollary is that many nations wish to have the opportunity to achieve the benefits of space on an essentially level playing field, while avoiding the potential deleterious uses of space by others. Accordingly, a level or even beneficial playing field becomes a broad international objective in terms of both benefits and threats. This chapter attempts to identify the forces tilting this conceptual playing field and to indicate the forces by which the playing field may be made at least acceptably level.

It is perhaps instructive to examine the role of treaties in the development of space. In a sense, the initial treaties, such as "The Treaty on Principles Governing the Activities of States in the Exploration and Use of Outer Space" have been designed to remove the economic or colonial motivation to develop space capabilities in respect to space exploration, and to protect the space assets established by the initial parties in space. Perhaps coincidentally, the initial space powers have also acted to protect the duopoly which they had at first achieved.

---

* This chapter is an edited version of a presentation made to the Fifteenth Annual Ottawa NACD Verification Symposium, "Arms Control and the Rule of Law: A Framework for Peace and Security in Outer Space," 11–14 March 1998, Montebello, Quebec, Canada.

1. This chapter is neither a scientific treatise nor a legally rigorous document. It is, rather, a series of views of a scientist, exposed to some aspects of the technical, commercial, and military applications of space, and under the benign influence of lawyers and politicians. It attempts to develop the position of an intermediate nation which, desiring the benefits of space, is perhaps realistic as to its participation in space, and, more significantly, the impact of the activities of others.

In parallel to this, the major powers have developed bilateral agreements with respect to arms control and disarmament which involve the secondary protection of national technical means (NTMs). Thus, the interests of the smaller powers have been largely ignored, except as they have reflected the interests of the emerging space powers.

The Space Context

Space as an endeavour and regime is unique in the history of humanity, but also manifests contrasts. Although it has not become a battleground to date, it has been an arena purporting to test and illustrate the advantages of alternative modern political ideologies. It has been a technological frontier providing a breakthrough to urge humanity in the exploration of other celestial bodies, and a vehicle for exploration of the vagaries of our own bodies as they react to lack of gravitational forces. It has proven to be a remote environment for the investigation of our Earth to new degrees of precision. In the realm of communications, it has brought humanity together as "a small village," while offering additional capabilities for "tribal but worldwide" warfare. It has revolutionized military doctrine, yet has not thus far been weaponized.

Above all, in relatively few decades space has become deeply pervasive in much of human endeavour, contributing to the military, economic, and social needs of many nations. As a consequence, space has acquired an aura symbolizing the status of a nation in regards to technological leadership and prestige amongst other nations.

National Capabilities

Space benefits, either civil or military, are achieved through a mix of highly advanced technologies — semi-conductors and electronics, high energy fuels and their engines, advanced composite materials, and computing, to name but a few. These must be coupled with detailed management, an industrial base, and the economic power to apply them. Thus, the benefits are not necessarily readily or equally available to all nations. Accordingly, we can classify nations by their space capabilities.

The two nations which have pioneered space and have the most highly developed capabilities (designated as Group I)[2] have largely attempted to restrain the "proliferation" of space technologies, while simultaneously moving themselves into strong positions militarily, technologically, and economically. It is, of course, difficult as well as risky to ascribe motives, but it is apparent that this duopoly provided not only rapid advances militarily: indigenously developed technologies, generally restricted to themselves; initially largely controlled access to space, allowing it to develop at their pace; and, protected market opportunities for their own industries. It is also possible to

---

2. Prevention of an Arms Race In Outer Space-United Nations A/48/305 15 Oct. 1993.

conclude that they have used their position to dominate the political/legal fora of space. Even the decline of the Soviet Union has not negated Russia's continuing ability to mount major space missions and to maintain a prominent position in political/legal discussions. The presently observed determination to maintain as much as possible of the duopoly must be, and is, of concern in legal fora. The superpowers have well-developed space capabilities, with the ability to independently mount virtually any missions deemed as necessary. It must be recognized that economics now also plays a significant part, to a degree not so clearly evident during the Cold War.

Close behind these leaders are the several nations which have demonstrated the autonomous ability to undertake specialized or, in a few cases, most space missions (designated as Group II). Into this group fall France, China, and Japan; candidates for inclusion in this group are Israel and India. There are other nations which possess long-range missile capabilities readily adapted to space and potentially able to use or refine them as a basis for future inclusion in this group. Nor should we neglect the potential of groups of nations collectively to achieve higher levels of self-sufficiency. In a sense, this is the impact of the Western European Union, NATO, and the European Space Agency (ESA), among others — although, as is discussed below, there is further significance to these structures.

The largest group is comprised of nations which have significant space capabilities in narrower areas, but are dependent, in one or more critical areas, on other nations to achieve the benefits of space (designated as Group III). Canada is a clear example, having space science interests, strong space communications and remote sensing capabilities, a good technological and industrial base, and even an astronaut programme. These are coupled with the demonstrated ability to design and build whole spacecraft. By policy design, however, Canada did not develop a launch capability, and is dependent upon the agreement of others to fulfil any space mission.

However, we note that the Canadian sounding rocket programme has the potential to move toward a small launch capability and, indeed, this has been proposed. As indicated, other nations within this group also would have the ability to move into autonomous space missions based on advances in their present rocket technology, coupled, in some cases, with increased payload and spacecraft capabilities. Clearly, those nations with a developed long-range or intercontinental missile capability are in a position to develop a launch capability.

In addition to these three categories, there are many nations which, although selectively participating in space systems and even owning part of a system (either ground equipment or a share of the space segment, i.e., INTELSAT) have little space technological capability in the sense of designing and implementing new hardware (designated Group IV). These are the nations which are perhaps most concerned with the economic and military significance of space. They tend to be unaligned with those nations having more extensive space capabilities, and their military forces are unlikely to be based on, or even to incorporate, significant space systems. Unfortunately many of these nations are also likely to be susceptible to economic exploitation by more advanced capabilities. These are the nations which are, or perhaps should be, most concerned about space, but also the least likely to be a threat to world stability based on space capabilities. However, we must not discount their potential to disrupt through non-space activities, to acquire limited aggressive space capabilities, or even be the focal point of a destabilizing space activity.

Speculatively, the threat in regard to major conflicts involving space seems to be in the lower capability end of Group II or Group III, while that of terrorism and lesser-scale conflicts would seem to be most likely in the bottom of Group III or the top of Group IV.

The final group (Group V), consists of the ever-diminishing number of nations which are not involved in space in some way.

Trends and Context

The interest in the grouping of the nations by their capabilities arises because they tend to cluster in their reaction to the general environment and trends, as well as being impacted in similar ways.

Since the end of the Cold War, there has been a very broad dispersal of space technology (not always deliberate or encouraged) across national boundaries — although, as noted, it has not necessarily resulted in the capability as much as in the desire to participate in the benefits of space. As an immediate consequence there are many more nations and/or commercial interests providing launch, satellite, or space services.

This dispersal of technological capability has arisen not only through the reduction of military tensions, but via the simple evolution of indigenous technologies with time and some technology transfer as a result of foreign involvement in civil programmes, coupled with the change of space from a military environment to a commercial market.

The rapid increase in the capabilities of commercial systems (in all space active Groups) has generated a situation wherein the performance of many commercial systems has reached the military requirements of a few years ago. These and even earlier military capabilities, particularly in remote imaging, were adequate as NTMs to provide initial verification of the present substantial arms reductions. Thus, today, civil space systems from several sources can provide at low cost and frequent coverage the data for many or most tactical and strategic military decisions. This duality of application by military and civil systems is perhaps the most significant trend of the present decade, and must be a significant driver in the future of treaties and agreements.

The projection that duality of system application is critical in respect to multilateral treaties is based not on the military use of civil systems *per se* as much as on the belief that over the next couple of decades space will become absolutely critical to the civil infrastructure of a substantial number of nations. Thus, in times of national emergency, it may be virtually impossible to withdraw or withhold space capabilities for military security without impacting one's own (and many other) civil operations.

The relaxation of Cold War tensions has also resulted in broader military alliances, with the result that more nations are incorporating space operations into their military doctrines. This has been accentuated by the demonstrations in recent conflicts — such as Desert Storm — of the effectiveness of space as a force multiplier.

Parallelling the broader military alliances generally involving an increasing number of nations in space, there have been strong forces leading to consolidation of commercial interests. Both within nations and across national boundaries there have been company acquisitions and alliances with attendant problems and advantages. The recognition that space offers unique capabilities in certain circumstances has resulted in space being incorporated into many civil systems. Not only has space provided low cost transoceanic and worldwide communications, as well as distribution and direct to home wideband entertainment, but also cost-effective imagery of the Earth as seen from orbit for resource assessments and exploitation, land use planning, precision navigation over the entire Earth, and so forth. This is a critical trend as it is impacting significantly the future of military operations.

Reflecting these shifts from the military/security concerns to the applications of space to the civil sectors, the space market has shifted from a military to a commercial/civil dominance. This seems to be reflected in the recent actions of both superpowers.

All these trends have become strongly established over the past decade, and it may be anticipated that they will continue over the one to come. It is clear that the current space treaties and agreements were not developed in the present context and are likely to become even less relevant as the trends continue, unless they are "upgraded" or replaced.

Military Space

In order to postulate an environment for future space agreements, it is useful to attempt an assessment of potential military space doctrine and how it might be best incorporated into treaties. It seems apparent that nations with significant space capabilities have incorporated these into their militaries and, although they may develop the ability to selectively destroy spacecraft or space transiting vehicles, they are unlikely to disrupt space operations.

It would seem that nations without general space capabilities are unlikely to have the specific abilities to selectively disrupt space segments. However, they may well have or acquire the ability to generally disrupt space systems if that becomes desirable in their estimations — a point to which we will return. Baines has provided a matrix approach providing weapon classes versus what he calls engagement scenarios, which provides guidance and projections as to space treaties (See Table I).[3] To this we add some general science/military considerations.

Table I. The Multilateral Arms Control Matrix

DEPLOYMENT LIFE-CYCLE	ENGAGEMENT SCENARIO			
WEAPON CLASS	Terre -to- Terre	Terre -to- Space	Space -to- Space	Space -to- Terre
WMD	a	b	X[c]	Xc
Conventional			d	d
"Exotic"			d	d

Key:

X Complete prohibition of a weapon class by multilateral agreement with substantial participation.

a No entry as CWC and BTWC only agreed to thus far. Nuclear element of WMD is still outstanding for weapons that "transits through" outer space.

b No entry as CWC and BTWC only agreed to thus far to prohibit deployment of WMD weapons that "reaches into" outer space. Nuclear component is still outstanding.

c The OST, LTBT, CTBT, CWC, BTWC collectively prohibit deployment of WMD that are space-based. Space-based means objects that complete at least one full orbit of any celestial body including the Earth, or on trajectories to or from any celestial body, or stationed in outer space in any other manner.

d Potential entries as a result of a Convention for the "Non-Weaponization" of Outer Space.

Space to Earth

In the case of space weaponization targeting Earth, it is suggested that there are several limitations which would make this class of weapon unattractive to a major space power. It is noteworthy that:

- most space segments spend their operational life in constant orbits which allows their tracking, prediction, and targeting;
- depending on the target, there are potential delays of an orbital period or more in availability of a space-based weapon. The most appropriate orbits

3. Phillip Baines, "A Variant for an Ad Hoc Committee on Outer Space Within the Conference on Disarmament," in *Final Report on 1998 Canadian Outer Space Workshop*, February 1998.

involve potential delays in targeting of the period of an orbit (90 minutes minimum) and, depending on the precise orbit and off-track capabilities, a delay of up to several days;

- the spacecraft design is indicative of the mission and data regarding the configuration can be gained by sensors on Earth as well as by other spacecraft;
- the orbital characteristics are highly indicative of the mission and targeting; and,
- most advanced weapons systems require maintenance which is difficult in low Earth orbit, not considered in higher orbits, and expensive in any case.

Finally, it is apparent that, in contrast to a direct missile attack, the total energy consumption, hardware requirements, and operational complexity involved in stationing a reentry type weapon in space, coupled with the reliability and vulnerability on station, make this a poor military choice which is not cost-effective. Notwithstanding the fact that such a weapon is not efficient, the possibility exists for a short-term blackmail scenario to be used by terrorists or a "rogue" nation.

Space to Space

If we initially restrict ourselves to weapons and targets stationed in space, we are dealing with a specific class of ASATs. In the major arms reduction agreements of the duopoly, there are clauses protecting NTMs. While these are clearly applicable to national military optical and radar spacecraft, perhaps as well as some specialized missions, it would appear that the relay and communications spacecraft involved in the NTM operations would also be protected. It would not be a vast stretch to include ELINT spacecraft as indicative of force size (even without the specific intelligence gained). In fact, it might be suggested that only the clearly civil spacecraft of the duopoly would be open legally to attack of any kind, while all the spacecraft of any other nations are vulnerable to attack. If the attack is predicated on the use of a nuclear weapon typified as being of "mass destruction," then such usage in space is prohibited. However, except for a short gain by a "rogue" nation without space assets, a multiple satellite kill scenario does not seem to be militarily credible. The alternative of "surgical" or selective destruction of satellites is a very real possibility, but is dependant on a relatively high degree of space capability in which case the aggressor military doctrine probably has a substantial sensitivity to the potential loss of its own space assets. It may be likened to a MAD of space assets and capability.

While the potential gain through attack of space systems in general does not seem to be acceptable, it is important to note that the value of assets in space is ever-increasing and that spacecraft, such as a space station, represent perhaps billions of dollars invested as well as multiple human lives, not to mention the social or public impact of an incident. Further, the space station orbits tend to traverse most of the world at relatively accessible altitudes. Thus, these assets may be put at risk by terrorism or "rogue" nations for short-term gain.

If the scenario shifts to space weapons versus vehicles transiting space or about to do so, the circumstances are quite different. First, space stationing of the weapons does make sense because it allows both observation and targeting to occur during boost phase when signatures are most intense and the consequences of destruction most benign to the attacker. It clearly requires a substantial number of weapons to be stationed in space if full launch time coverage of all vehicles is to be achieved. This is, of course, the ABM regime. It probably is solely the interest of the Group I nations or the superpowers.

The case of the Earth-to-space class of ASATs is not really a part of outer space agreements, although it is clearly a real potential threat to space assets. It does, of course, come under the non-interference with NTMs of the duopoly. The ability to attack space selectively from the ground does not necessarily involve space capability *per se*, but does imply considerable high technology in the exotica such as beam weapons. The sole saving grace is that without weapons in space it is difficult to achieve a simultaneous kill. Thus, a surgical orbit cleansing can be achieved, but it takes an extended period which is likely to result in rapid retaliation before it is completed.

Space as a Playing Field

The majority of nations would like guarantees that space will not be used against them and that they will be able to use it to the maximum extent possible. Thus, most nations desire a level playing field. However, the field is subject to tilts by virtue of space being the subject of national prestige, national security, technological capability, market protectionism, internal and international commercial competition, available financial resources, and political interests. While most of these factors are self-evident, some are expanded upon below.

Procurement Barriers and Technology Transfer

Space is unique in the commercial sense, in that the customer requiring a space asset can rarely contract for a fully protected failure situation in the sense of being guaranteed the full cost recovery or, optimally, the profit of the system. This is exacerbated by the lead time for a typical space system, which is rarely "off the shelf," and this results in an extended down time for a system in the event of a failure.

Inasmuch as the risk becomes a shared one, the customer usually requires an extremely high degree of visibility regarding the design and fabrication operations of the supplier. This introduces a major cost item but, more significantly in the present context, will require a substantial indigenous (or contracted) technical ability and up-to-date knowledge as to the potential of the "state-of-the-art," often commensurate with, or bettering, that of the supplier. Thus, there are immediate concerns based on the direct or potential technology transfer which may develop.

In procurement crossing national boundaries this visibility demand may be hard to achieve. These concerns may, of course, be either of a proprietary nature or a military concern or both. Realistically, the military concerns expressed are often developed or enhanced by what are actually competitive considerations.

Customs barriers can also be a problem. However, at least in Canada, it has been possible to bring critical hardware into the country with no duty or with a recoverable hold-back on the grounds of being "re-exported" to outer space at a later date.[4] To date, there has been no invocation of the potential legal problem associated with "an entry into the commerce of the nation," despite being in outer space.

Access to Space

Fortunately, the proliferation of potential launchers, coupled with competitive commercial pressures has made it possible now to have a largely level playing field with regard to access to space. Even the military aspects of the launch vehicles realistically do not come into play to any great extent (although they may be called up) since the customer for launch services rarely has any great visibility into their technology beyond the interfaces. Conversely, the launch supplier may have a greater visibility into the technology of the space assets.

Under the present circumstances, it has been suggested that the commercial pressures/opportunities will ensure launches for all nations, but one might question this availability in the case of a clearly military spacecraft.

It has been suggested that the launch/spacecraft mate and launch operations provide a reasonable period (usually about a month) which could allow international inspections to determine the nature of a spacecraft with a view to eliminating weapons in space. This might be a key element if it were decided that nations be guaranteed access to space by treaty. Comprehensive launch/spacecraft inspection, particularly over the latter stages of integration and pre-launch assembly, can provide some degree of certainty with respect to weapons of mass destruction, several forms of space mines, and some configurations of satellite weapons. However, to generalize inspection across all launches — particularly those of nations having indigenous launch capabilities — is likely to prove very difficult. Here, one might invoke citizen-based monitoring capabilities to provide an inhibiting factor.[5]

---

4. One interesting consequence was that if a truck carrying certain major hardware for a spacecraft was in an accident or destroyed in Canada, the drivers were under orders that it was to be towed into the United States so as to complete the exportation and allow retrieval of several million dollars in duty holdbacks.
5. Peter Stibrany, "Alternative Methods for Compliance Monitoring," 1998 Outer Space

Treaties could guarantee access to space for both civil and military payloads which, in turn, might inhibit a race to develop autonomous launch capabilities. In that sense, the reduction in launch proliferation would be stabilizing, while the fact that launches would be provided under international or treaty controls would provide a dampening effect and allow additional influences to act against a race toward militarization of space. Obviously, in the present context, weaponization would be prohibited. Similarly, treaties which provide guarantees of security against aggression tend to reduce the need for excessive militarization of all forms, including in space.

There are various mechanisms which serve to reduce the tilt on the playing field. For example, military alliances tend to involve the smaller players as well as introduce space into their military doctrines. International programmes perform the same function as commercial or civil programmes, and provide a further mechanism for the dissemination of technology. A major factor in changing (either way) the tilt is the consolidation of domestic companies and their generation of alliances or ownership across national boundaries. Finally, international law and treaties can serve to flatten the field.

What treaties are unlikely to do is reduce the desire for a dominant nation to maintain its military advantage, or that of a somewhat lesser power to achieve a balancing capability where there is a sense of risk.

Thus, it is suggested that treaties will, in general, reduce the risk of an arms race in space, but will impact most significantly those nations of moderate capability. However, treaties will not significantly decrease the potential or desire for a single nation to give itself a significant military or weapons capability in space, particularly if that nation has an aggressive stance or intent and/or believes that in doing so it levels the military playing field. Thus, one must consider whether a nation without a space-based military doctrine might not leap at a chance to carry out an attack on selected, or all, space assets, or to be in a position to threaten to do so.

Conclusion

It is postulated in this chapter that a fully level playing field in terms of space capability, application, access, and benefits — including military, economic, and cultural — is unrealistic. Conversely, it is suggested that treaties in the nature of guarantees which tend to level access to and use of space tend to be stabilizing and increase the well-being of nations without detriment to others. In a sense, each nation is allowed to develop and/or use space to the maximum of its own potential or needs without being unduly limited by its neighbours or their policies, and without the necessity to develop inappropriate space capa-

---

Workshop. Please also see Stibrany, "The Role of Non-Governmental Organizations in Maintaining Space as a Weapons-Free Environment," chapter eleven of this volume.

bilities — i.e., to indulge in a space race of some kind other than for its own well-being.

It is important to note that treaties can provide these benefits by ensuring equality of access and utilization in any manner. They need not guarantee a national space launch or satellite capability, but rather the benefits. For example, guarantees of access to national multi-spectral imagery at commercial rates provide the same benefits as building and launching a remote sensing satellite.

The military use of space is well established. Particularly, the observational capabilities can be considered as security-enhancing, and are rapidly becoming a cornerstone of world disarmament. They are likely to remain so and are highly complemented by the civil capabilities. However, the weaponization of space is believed not to have developed yet, nor is space thought of as being militarized (in the sense that the military control access to it). While there have been legal agreements tending to inhibit the weaponization of space, there are also valid technical reasons for the willingness to forego stationing weapons in space.

If we consider a weapon designed for "orbital cleansing" — i.e., an attack by a single or multiple weapons on a large volume of space — it is apparent that it is most likely to originate with a party without major space assets which it would hope to maintain. Thus, it is most likely to be a non-space-based military force trying to level the playing field by means of a single launch, an effort to survive a losing situation, or again a terrorist group or nation.

It appears at the present time that space has not been weaponized, nor is any immediate threat apparent. However, it is also clear that the present space treaties do not, either in terms or scope of signatories, reflect the present significance of space in regard to both military and civil sectors, and are inadequate to ensure that space will remain free of weapons. It appears that there is now a window of opportunity to develop new space treaties which can allow space to develop in the interests of all of humanity.

Chapter Two

The Relationship of the ABM Treaty to Strategic Stability and Outer Space

Patricia Bliss McFate and Sidney N. Graybeal*

The Negotiating Background of the ABM Treaty

After the Soviet Union's detonation of a nuclear device in August 1949, the United States, taking into account the intelligence community's projections of Soviet long-range bomber capabilities, embarked on the development of a strong air defence capability. As the US air defence capability was beginning to mature, two events took place that were considered additional indications of a new emerging Soviet threat: a successful Soviet intercontinental ballistic missile (ICBM) test in August 1957 and the launch of *Sputnik* in October of that year.

The United States and the Soviet Union developed ABM defence concepts and programmes concurrently with ICBMs. In view of the extensive Russian air defence system, serious questions arose regarding the ABM capabilities of several surface-to-air missile (SAM) systems. The "SAM-upgrade" question was a serious concern throughout the ABM Treaty negotiations, and it remains an issue related to US and Russian theatre missile defences.

It was clear by the early 1960s that a large-scale Soviet ABM programme was underway, a conclusion reinforced by General Secretary Nikita Khrushchev's claim, in 1962, that the Soviet Union could "hit a fly in outer space." In 1962–1963 construction began on a large ABM system at Moscow, utilizing a nuclear-armed ABM interceptor. Four ABM complexes, with a total of 64 launchers, were completed, and the system continues to be upgraded.

In the Kennedy and Johnson administrations, Secretary of Defense Robert McNamara's view of strategic deterrence was embodied in the concept of mutually assured destruction (MAD). Arms control was considered consistent with this strategy; deployment of extensive ballistic missile defence (BMD) was not. McNamara was not convinced that a BMD system was technically feasible and capable of defending the US population against the emerging Soviet ballistic missile threat.

---

* This chapter is an edited version of a presentation made to the Fifteenth Annual Ottawa NACD Verification Symposium, "Arms Control and the Rule of Law: A Framework for Peace and Security in Outer Space," 11–14 March 1998, Montebello, Quebec, Canada.

During the Glassboro, N.J. summit meeting between President Lyndon Johnson and Premier Aleksei Kosygin in June 1967, Secretary McNamara argued that there was a direct relationship between ABMs and strategic ballistic missiles that affected both crisis and strategic stability. When the United States raised the desirability of limiting ABM systems on the grounds of this link between strategic offensive and strategic defensive arms, Kosygin reportedly responded, "Why do you want to limit ABMs? They don't kill people; they protect them." A few months earlier, according to *Pravda*, Kosygin had stated: "I think that a defensive system which prevents attack is not a cause of the arms race. ... Perhaps an anti-missile system is more expensive than an offensive system; but its purpose is not to kill people but to save human lives." McNamara pointed out the potential destabilizing effects of nationwide ABM defences: with deployed effective nationwide BMD systems on both sides, there could be a temptation to shoot first in a time of crisis inasmuch as the initiator's offensive forces would be able to penetrate the other side's defences and its defences would be able to defend against the reduced, ragged retaliatory attack. McNamara also pointed out the likelihood of such defences contributing to an arms race in which both sides attempted to acquire more and better strategic offensive forces.

While Premier Kosygin had not been convinced of the desirability of limiting ABM systems at Glassboro, Soviet agreement to bilateral negotiations on limiting strategic offensive and defensive arms was announced by President Johnson on 1 July 1968, the day of the signing of the Non-Proliferation Treaty (NPT). The Soviet occupation of Czechoslovakia in August forced Johnson to postpone the negotiations, leaving the issues for President Richard Nixon's administration to settle. In October 1969, Nixon announced that he had agreed with the Soviet leaders to begin talks on strategic arms. The negotiations, known as the first round of the Strategic Arms Limitation Talks (SALT I), began in Helsinki, Finland, on 17 November 1969.

During the early rounds of SALT, it became clear that the Soviets were prepared to limit ABM systems. The basis for this change in the previous Soviet view was either recognition of the logic of McNamara's strategic stability arguments or a desire to curtail US ABM systems. The two sides proposed strict limits on ABM systems, but serious differences over offensive limitations emerged. These disagreements reflected major disparities in the makeup of the two sides' strategic forces, especially the Soviets' reliance on large land-based ICBMs and their desire to include in the accord the strategic forces held by US allies and US forward-based weapons on aircraft carriers and foreign bases.

Faced with the prospect of delays in negotiating progress due to disagreements over offensive limitations, the Soviet Union proposed that the sides agree on an ABM Treaty while leaving offensive forces to a subsequent negotiation. However, the United States insisted on a dual agreement, hoping to use its ongoing ABM deployment programme to gain Soviet concessions on offensive arms. In a back channel negotiation, Henry Kissinger, Nixon's national security adviser, and Anatoly Dobrynin, the Soviet ambassador to the United States, reached agreement in May 1971 that a comprehensive ABM accord would be accompanied by a more limited agreement on certain offensive arms, leaving more comprehensive offensive limitations to SALT II.

One year later, on 26 May 1972, President Nixon and Soviet leader Leonid Brezhnev signed the SALT I agreements, including both the ABM Treaty and the Interim Agreement on Limiting Strategic Offensive Arms. The ABM Treaty banned nationwide ABM systems, while allowing each side to maintain the ABM site it then possessed and, for the sake of symmetry, to build one of the type then being built by the other side. The interim agreement froze each side's ICBM and SLBM launchers at the level then operational or under construction for five years, but it did not place any limits on multiple independently targetable re-entry vehicles (MIRVs), bomber forces, or forward-based systems. The US Senate gave its advice and consent to ratification of the ABM Treaty on 3 August by a vote of 88 to 2, and both houses of Congress gave similarly overwhelming approval to the interim agreement. The accords entered into force on 3 October 1972.

The basic purpose of the ABM Treaty is contained in Article I, which prohibits deployment of an ABM system for "the defence of the territory" or provision of a "base for such a defence." The former prohibition refers to a nationwide defence; the latter encompasses items that could be the long-lead-time supporting elements for a deployed land-based ABM system — i.e., large phased-array radars (LPARs). All other treaty articles and agreed statements are designed to support Article I.

Early in the SALT negotiations, both sides recognized the need for an implementing body. There were no serious problems in negotiating Article XIII of the Treaty, which calls for the prompt establishment of a Standing Consultative Commission (SCC) to promote the objectives and implement the provisions of the Treaty. The charter of the SCC, the first of its kind in arms control agreements, is contained in a memorandum of understanding signed in Geneva on 21 December 1972; a protocol with regulations for the SCC was signed on 30 May 1973.[1]

Treaty Implementation and Operation

During its 25 years of operation, the SCC has successfully negotiated the dismantling or destruction procedures called for in the ABM Treaty and in the interim agreement, has dealt effectively with ambiguities and compliance issues, and has negotiated legally binding clarifications, modifications, and

---

1. A coauthor of this paper, Sidney N. Graybeal, who negotiated Article XIII and the SCC charter and regulations, was appointed by President Nixon to serve as the first US SCC Commissioner in June 1973. He served in this capacity through 1976.

amendments to the Treaty. To date, numerous compliance questions and ambiguities have been raised and clarified in the SCC.

The Soviets, and now the Russians, have continued an active missile defence research and development programme since the Treaty was signed, including traditional ABM components and new technologies, such as lasers.

US consideration of a national missile defence (NMD) took a dramatic turn with the "Star Wars" speech of President Ronald Reagan on 23 March 1983. In the speech, the president announced a proposal to develop an anti-missile defence system so comprehensive that it would render "nuclear weapons impotent and obsolete." Reactions to the proposal ranged widely: some condemned it as impossible and destabilizing, while others saw it as the potential saviour of the nation. While senior officials supported the proposal, in large part because it represented a personal commitment on the part of the commander-in-chief, Richard DeLauer, Pentagon chief of research and development stated, "There's no way an enemy can't overwhelm your defenses if he wants to badly enough."[2] In January 1984, the Strategic Defense Initiative Organization (SDIO) was created to undertake a programme to develop technologies associated with ballistic missile defence.

The US-Soviet Nuclear and Space Talks (NST), which commenced on 12 March 1985, incorporated discussions of strategic offensive and strategic defensive systems and their relationship. There were major areas of substantive disagreement in the Defense and Space forum of the NST. The Soviets contended that an agreement concerning compliance with the ABM Treaty was a condition for a reduction in strategic offensive arms; the United States opposed the linkage.

On 11 October 1985, President Reagan announced his determination that a broader interpretation of the ABM Treaty was fully justified. Under the broad interpretation, or the so-called reinterpretation of the treaty, ABM systems that are based on "other physical principles" (i.e., other than ABM interceptor missiles, ABM launchers, and ABM radars known in 1972) and including components capable of substituting for ABM interceptor missiles, ABM launchers, or ABM radars may be developed and tested but not deployed, regardless of their basing mode. Under the narrower, traditional interpretation, development and testing of ABM systems based on other physical principles — for example, lasers — are allowed only for fixed land-based systems and components. Immediately after the announcement, Secretary of State George Shultz explained that, while the new interpretation reflected US legal rights, the United States would not change its policy of conducting the SDI programme according to the narrow interpretation of the Treaty.

The most effective opponent of the reinterpretation was the United States Congress, in particular the Senate. The lead role of challenging and eventually

---

2. Arms Control Association, *Star War Quotes* (July 1986), 34.

defeating the reinterpretation was undertaken by Senator Sam Nunn, then chairman of the Senate Armed Services Committee. Senator Nunn's review of the ratification record found that the Nixon administration presented the Senate with the historic interpretation of the Treaty's limits on space-based and other mobile types of exotic systems, and the Senate clearly understood this interpretation when it gave its advice and consent to the ratification of the Treaty. The immediate result of congressional review was to block the administration's switch to the broad interpretation. In the Defense Authorization Acts of fiscal years 1988 and 1989, Congress enacted provisions requiring that SDI funds be spent only on activities consistent with the narrow interpretation. By this time, even Nixon, who had signed the ABM Treaty in 1972, had stated publicly his unequivocal support for the historic interpretation.[3]

The United States tabled a draft treaty in the NST in 1988 calling for research, development, and testing of advanced defences based on the "broad interpretation" of the Treaty; the Soviets rejected the interpretation. The United States called for its right to withdraw from the Treaty after a specified period of time; the Soviets argued that Article XV of the Treaty did not permit withdrawal simply in view of the successful development of an effective strategic defensive programme. While the United States proposed making a transition from deterrence based on the threat of nuclear retaliation (mutually assured destruction) toward increased reliance on defences against ballistic missiles (mutually assured survival), the Soviets sought a comprehensive ban on research, development, testing, and deployment of "space-strike arms."[4]

The Reykjavik summit meeting on 11–12 October 1986 between President Reagan and General Secretary Gorbachev came close to an agreement on significant reductions in US and Soviet strategic offensive missiles. The Soviets proposed that the United States provide a ten-year commitment not to withdraw from the ABM Treaty. The United States offered to accept the commitment through 1996, provided research, development, and testing in accordance with the broad interpretation could continue. The Soviet response, which was unacceptable to Reagan, was a proposal to ban testing of space-based elements of a missile defence system outside of laboratories. Thus, the summit failed to make progress on either reduction of strategic offensive arms or ABM Treaty clarifications.

The ultimate goal of the SDI programme was complete US homeland protection from a massive Soviet nuclear attack. But the plan was to be developed incrementally. The initial step, Phase One, was designed to blunt,

---

3. Matthew Bunn, "Nixon Supports Nunn on ABM Treaty," *Arms Control Today* (July 1988), 21.

4. Discussion of the US negotiating position is based on the Arms Control and Disarmament Agency's *Annual Report to Congress, 1988*. Discussion of the Soviet position is based on a translation of the article, "Scientific Fact Must Back Political View of Disarmament," by Aleksey G. Arbatov in *Mirovaya Ekonomika* (April 1988), 10–22.

not completely stop, a massive attack. One of the key elements of the defensive system was the use of space-based interceptors and sensors. At first, SDIO considered individual space-based interceptors but found their cost prohibitive. Then came the idea to place up to ten space-based interceptors into "space garages" that would cut costs but provided an attractive target. The next step was called Brilliant Pebbles, envisioned as a system of low-cost, mass-produced, highly autonomous, and individually deployed space-based interceptors. However, the Brilliant Pebbles programme never lived up to the promises of its advocates, and Congress had deep concerns about weapons in space and a programme that was clearly noncompliant with the ABM Treaty.

As time passed, the initial enthusiastic endorsements for the SDI diluted as the scientific and technological community weighed in with their skeptical responses. In January 1988, Senator Nunn proposed that near-term SDI research be redirected toward exploring the feasibility of an Accidental Launch Protection System (ALPS), a very limited treaty-compliant system designed to defend against a few missiles launched by accident or without authorization. Nunn's proposal touched off a wide-ranging discussion of whether accidental or unauthorized attacks represented a significant threat and of the pros and cons of the ALPS approach to dealing with them.

In the Missile Defense Act of 1991, Congress went on record as supporting deployment of a US missile defence. The act endorsed the concept of a limited defence "designed to protect the United States against limited ballistic missile threats, including accidental or unauthorized launches or Third World attacks." One of the key provisions was a mandate to develop an NMD by 1996 or as soon as the technology was ready. The single US defensive site was to contain 100 ground-based interceptors and be fully compliant with the ABM Treaty. This approach was completely consistent with Senator Nunn's ALPS proposal of 1988.

On 17 June 1992, Presidents Bush and Yeltsin signed a joint statement in which they agreed to cooperate in the development of a concept for a Global Protection System (GPS). The Russians used the term Global Defense System; they believed that this system could be accomplished outside of, but consistent with, the ABM Treaty. At the Washington summit, Bush and Yeltsin agreed to "explore on a priority basis" potential opportunities for sharing data (primarily from space-based systems) in a ballistic missile early warning system, for cooperating in ballistic missile defence capabilities and technologies, and for developing a legal basis for the implementation of GPS, including the possible negotiation of new treaties.

Although US theatre missile defence (TMD) systems are receiving the highest priority, the Department of Defense is also developing for deployment a ground-based, treaty-compliant NMD system, known as the "3+3" system (three years for development and three years until deployment) as a hedge against the possibility that new strategic ballistic missile threats may emerge after the turn of the century. The threat of an intercontinental ballistic missile attack on the United States has markedly changed. The Cold War threat — an all-out, high-level attack from the Soviet Union — has disappeared. Limited ballistic missile threats to the continental United States from accidental or unauthorized Chinese or Russian launches are considered unlikely in the foreseeable future. However, some Third World countries could acquire long-range ballistic missile capabilities within the next ten years. As former US director of central intelligence James Woolsey has noted, "After the turn of the century, some countries that are hostile to the United States might be able to acquire ballistic missiles that could threaten the Continental United States ... over the next ten years we are likely to see several Third World countries at least establish the infrastructure and develop the technical knowledge that is necessary to undertake ICBM and space-launch vehicle development."[5]

The issue of the compliance of space-based sensors continues to be raised, both within the government and in the arms control community outside the government. Congress clearly stated its belief that space-based sensors are not prohibited by the Treaty in the Missile Defense Act of 1991, as modified by the fiscal year 1995 DoD Authorization Act, when it indicated that there should be "optimum utilization of space-based sensors, including sensors capable of cuing ground-based antiballistic missile interceptors and providing initial targeting vectors, and other sensor systems that are not prohibited by the ABM Treaty."

While there are limits on ABM radars in the Treaty, the only constraints on non-ABM radars or other sensors are placed on the deployment of early warning LPARs — they must be located along the periphery of the country and oriented outward. There are no limits on the amount and quality of data that may be transmitted to a battle management centre, and there are no treaty limits on battle management. Thus, data from early warning radars can be transmitted to NORAD, just as data from LPARs can be transmitted to Moscow.

This position regarding LPAR data and its use was confirmed in the Bush Administration's Compliance Report to Congress dated 19 January 1993. The report stated that the US Government:

> ... now judges that the support of ABM systems by early warning radars providing precise handover data will not constitute use of the early warning radars as ABM radars in violation of the ABM Treaty. Specifically, [the US government] will not consider as prohibited the handover of precise target state vectors by properly located and oriented early warning radars to ABM systems or ABM components.[6]

---

5. Testimony before the Senate Governmental Affairs Committee on 24 February 1993.
6. "Adherence to and Compliance with Arms Control Agreements, and the President's Report to Congress on Soviet Noncompliance with Arms Control Agreements," US Arms Control and Disarmament Agency, transmitted to the US Congress by President George Bush on 19 January 1993, 11.

The report concludes that "handover by an early warning radar to an ABM system or ABM component would not constitute testing in an ABM mode," nor giving of "capability to counter strategic ballistic missiles," which are the limits provided in Article VI (a) of the Treaty. This position was accepted by the Clinton Administration.

The foregoing conclusions on ground-based sensors should also apply to data from space-based sensors, as well as sea-based and air-based sensors. All these external sensors can transmit data to battle management centres, which can then be used to enhance the capabilities of an ABM system completely consistent with the Treaty. Neither space-, air-, sea-, or land-based sensors, nor the quantity or quality of the data they transmit, are limited by the Treaty, unless the data are transmitted directly to an ABM interceptor while in flight. In that case, the data could be providing guidance, thus making the sensor an ABM component prohibited by the Treaty.

To date, the implementation and operation of the Treaty has been extremely effective in the eyes of most national security experts. They believe it remains in the US national security interest. It is amazing that a treaty which is over twenty-five years old has fared so well. It was intended to be a living document — to be clarified, modified, and amended as required. This has, and is, happening. However, some argue that it still requires resuscitation to remain consistent with US national security objectives, while others maintain that it is a relic of the Cold War which interferes with the ability of the United States to acquire effective TMD and NMD systems and that it should be abandoned.

The Future of the Treaty

The preamble of the ABM Treaty contains it basic premise: "Effective measures to limit anti-ballistic missile systems would be a substantial factor in curbing the race in strategic offensive arms and would lead to a decrease in the risk of outbreak of war involving nuclear weapons." This premise remains valid, at least through the implementation of the START agreements, when substantially lower levels of strategic forces should lessen the risk of war, as well. The offense-defence link and concerns about strategic and arms race stability will continue into the next century; indeed Russia has specifically linked START limitations to maintaining the viability of the Treaty.

In 1972, the year in which the ABM Treaty entered into force, the United States possessed approximately 4,000 warheads on some 1,700 strategic ballistic missiles; the Soviet Union had approximately 2,050 warheads on about the same number of strategic ballistic missiles. Even when the START II limits are reached in 2003,[7] each side will still be able to retain 3,000–3,500

7. An agreement between Presidents Clinton and Yeltsin signed on 28 September 1994 calls for an expedited deactivation process once START II is ratified. Subsequently,

strategic warheads, a significant number of which will remain on strategic ballistic missiles. In 1997, the leaders of Russia and the United States committed to the prompt negotiation of a follow-on treaty, START III, as soon as START II is ratified. START III would lower deployed weapons levels by another 1,000 warheads by the end of 2007.

The Clinton Administration is committed to maintaining the integrity and viability of the Treaty, although it recognizes the continuing requirement for further clarifications, modifications, and amendments. While the Administration believes that the Treaty remains in the national security interest of the United States, it also states that it must be clarified and amended in order to achieve required, effective TMD systems. Since 1972, ballistic missile threats have changed. Advanced technologies and systems are becoming available for the development and deployment of both theatre ballistic missiles and BMD systems. The ABM Treaty must take into account these emerging threats and advancing technologies.

Article VI (a) of the Treaty prohibits giving non-ABM components and systems "capabilities to counter strategic ballistic missiles or their elements in flight trajectory." This is the "SAM-upgrade problem" referred to earlier in this chapter. With the advanced TMD systems being developed in the United States and Russia, distinguishing between these TMD systems and ABM systems became increasingly difficult. Consequently, the SCC negotiated two accords which address demarcation: they establish criteria for low-velocity and high-velocity TMD interceptors, in order to distinguish them from ABM (strategic) interceptors. Another accord, which deals with the successor states issue, clarifies that Belarus, Kazakhstan, and Ukraine will join Russia in becoming full Treaty parties. The demarcation and the multilateralization agreements must be submitted to the US Senate for its advice and consent.

This upcoming ratification debate will open up many additional ABM Treaty issues, including whether or not the Treaty remains in the US national security interest. Senator Jesse Helms, Chairman of the Senate Committee on Foreign Relations, has already set forth his views:

> The Senate has not had an opportunity to consider the rationale behind the ABM Treaty since that treaty was ratified nearly 26 years ago, in the midst of the Cold War. The world has changed a great deal since then. It is vital that the Senate conduct a thorough review of the ABM Treaty this year when it considers and votes on the ABM Multilateralization and Demarcation agreements.[8]

---

in 1997, in part to help with the Russian Duma's ratification of START II, the United States and Russia agreed to shift the Treaty's final implementation date by five years — from 1 January 2003, to 31 December 2007. The two countries also reached agreement in principle on amending START I to make its duration indefinite.

8. Letter from Senator Helms to President Bill Clinton dated 21 January 1998.

As technologies develop, issues associated with demarcation will increase. For example, the United States has an ongoing airborne laser (ABL) programme which involves installation of a multi-megawatt laser aboard a 747-400 freighter aircraft. The ABL is designed to destroy missiles launched hundreds of kilometres away. The laser heats a basketball-sized circle on the outside wall of a missile casing, causing it to rupture during the boost-phase. Although this system is designed primarily for a TMD role, it would clearly be capable of destroying ICBMs and SLBMs in their boost phase, depending upon the location of the aircraft.

While consensus on the emerging *theatre* ballistic missile threat has developed in the United States, making the case for a *strategic* ballistic missile threat to the United States is far more controversial. Some would argue that the threat exists today in the form of a possible accidental or unauthorized launch from Russia or China, and the possibility of a "rogue" state acquiring a space launch vehicle which can be readily converted into an ICBM. Others maintain that there is no current ballistic missile threat to the United States, and that there will be sufficient intelligence warning of such an emerging threat in sufficient time to deploy an NMD system capable of defending against this limited threat. The Clinton Administration's "3+3" NMD programme mentioned above is designed to meet this threat.

Some Members of Congress and a few non-governmental organizations and individuals are urging the development and deployment of a space-based laser system which would be capable of defending against theatre ballistic missiles, SLBMs, and ICBMs. While laboratory research and development of such a system is not prohibited by the ABM Treaty, the development and testing of a prototype would be prohibited. This group also notes that such a space-based system would be capable of providing the United States with control of space, which they maintain is essential for current and future US war fighting and intelligence capabilities. These individuals are well aware that the ABM Treaty prohibits space-based ABM systems. Thus, when they maintain that the Treaty is interfering with effective BMD systems, they also argue that the United States should withdraw from the Treaty.

An alternative NMD system proposed by a study team associated with The Heritage Foundation, involves utilizing a fully-upgraded shipborne weapons system called *Aegis* which could be used for sea-based defence against long-range and short-range ballistic missiles.[9] The system would utilize data from Brilliant Eyes, a sensor satellite capable of detecting missile launches and tracking the missiles in flight. The group proposes that the system be developed and deployed without any ABM Treaty restraints. Since this system would be *both* a TMD and an NMD system it would violate Article VI (a).

---

9. *Defending America: A Near- and Long-Term Plan to Deploy Missile Defenses*, a report by the Missile Defense Study Team, The Heritage Foundation, 1995, 4.

On the other side of the argument are a significant number of Members of Congress and non-governmental organizations and individuals who strongly support maintaining the basic purpose and intent of the ABM Treaty. They argue that the Treaty is necessary for further reductions in strategic offensive arms, for maintaining strategic and arms race stability, and for continued implementation of Article VI of the Non-Proliferation Treaty.

The Russians have remained consistently positive in their view of the ABM Treaty. Russian Defense Minister Igor Sergeyev recently stated: "It is impossible to evade the ABM Treaty; otherwise we will again fall into coils of the arms race and it will be hard to speak of further arms reductions."[10]

In the late 1990s, a repeated American refrain heard from the US military, and now echoed by certain Members of Congress, has been "space dependence," leading to a requirement of "space control." As noted in a recent study by the National Defense Panel, US assets in space are vulnerable and lack the ability to detect attack — protection and denial capabilities are rudimentary, and are limited to encryption of communication links, some degree of hardened electronics, and enough redundancy to guard against catastrophic loss of capabilities. "Denial of enemy space capabilities is largely limited to neutralizing enemy ground installations employing convention or special operations forces." The Panel recommended that the United States "take steps now to ensure we have the capability to deny our enemies the use of space."[11]

Anti-satellite (ASAT) systems *per se* are not limited by the ABM Treaty. This was a conscious US decision during the negotiation of the Treaty. However, Article VI (a) precludes the testing of any ASAT system against a strategic ballistic missile or its elements in flight trajectory. Furthermore, Article XII of the Treaty provides that it be verified solely by national technical means (NTM).[12] A unique aspect of the ABM Treaty verification regime was that it prohibited each party from interfering with the NTM of the other party. Inasmuch as some reconnaissance satellites are an essential part of NTM, it would appear that developing and deploying any ASAT system would be counter to the purpose of this article. In addition, deployment of some currently-proposed ASAT systems, such as ground-based and space-based lasers which have significant ABM capabilities, would be a violation of the Treaty.

On 1 September 1997, an article by William J. Broad in *The New York Times* announced that the US military was seeking permission to fire the nation's most powerful military beam into space at an Air Force satellite. The laser utilized in the test was the Mid-Infrared Advanced Chemical Laser

---

10. Quoted in *BMD Monitor*, 19 September 1997.
11. *Transforming Defense: National Security in the 21st Century*, Report of the National Defense Panel, December 1997, 38–39.
12. National technical means include reconnaissance satellite systems using a variety of sensors, ground- and sea-based radars, seismographs, communication collection stations, and underwater acoustic systems.

(MIRACL), a ground-based megawatt class deuterium-fluoride laser operating in a band from 3.6 to 4.2 microns; it is the highest average power laser in the United States. As Broad noted, the test "would be a major step toward perfecting a weapon that could demolish satellites and other spacecraft, an ability that the Pentagon regards as crucial in time of war." The test was delayed, in part, because several Democratic Members of Congress wrote President Clinton stating that they were "deeply troubled that a test of a ground-based laser system with such obvious ASAT warfare capabilities would proceed ahead of any debate or deliberate policy development."[13]

In mid-September 1997, Russian President Yeltsin wrote to President Clinton stating that the Russians were "alarmed by the US military's intention to develop a whole gamut of anti-satellite weapons systems."[14] He continued:

> The aim is obvious — to develop a capability to destroy space surveillance and control systems of other countries, including, of course, the Russian ones. That means that missile attack warning and surveillance systems that ensure the required transparency and therefore predictability and stability may be jeopardized.

President Yeltsin proposed negotiating a ban on any system capable of destroying strategic warning satellites, followed by joint work on the problem of "possible destabilization in low orbits." With the current US military emphasis on control of space, it is unlikely that serious negotiations on limiting ASAT systems will take place in the near future.

In a related event, the Conference on Disarmament (CD) in Geneva passed a resolution in 1997 which called for the CD to reexamine and update the mandate for establishing an ad hoc committee on preventing an arms race in outer space, citing "developments in laser technology and the US-driven pressure to weaken the ABM Treaty."[15]

Many countries have satellite programmes; the majority of these are for commercial purposes. The more developed countries blend defence, civil, and commercial satellite systems for military purposes as well. As William Scott recently noted, this blending "raises questions about international space policies, vulnerabilities, and what constitutes an act of war in space during times of heightened international tensions."[16] Recent war games developed for the

---

13. Letter to President Clinton from Richard A. Gephardt, John Spratt, Jr., and Ronald V. Delhums, Members of Congress, dated 26 September 1997.
14. Although not formally released by the Clinton Administration, the letter has been reproduced by several non-governmental organizations. These quotations from an unofficial translation come from a publication of the Center for Security Policy.
15. "Multilateral Arms Control: Can the CD Break the Impasse?" *Arms Control Today* (November/December 1997), 20.
16. "War Game Raises New Space Policy Dilemmas," *Aviation Week & Space Technology* (23 February 1998), 98.

US Army reveal possible trends in that those playing the adversary have held back from space war, apparently recognizing the implications of taking too strong an action in space.

Other positive activities include a joint commercial US-Russian space mission to obtain marketable high-resolution/map quality images of Earth, utilizing a Russian spacecraft previously used for Soviet defence mapping operations. In addition, Russian and American scientists have been working for three years on development of a space-based remote sensor called RAMOS. RAMOS is currently using US aircraft and US and Russian satellites to collect stereo imagery of backgrounds of interest in the visible and infrared wavelength. If bilateral relations continue to improve, the two nations could begin sharing data and cooperating in BMD exercises, capabilities, and technologies. To meet the future proliferation of advanced cruise missiles, the cooperation could also extend to advanced defences against aerodynamic vehicles.

The World With and Without the ABM Treaty: A Prospective Look

It is not known whether the ABM Treaty will exist in 2010. However, taking into consideration certain geopolitical constraints and realities, it is possible to identify certain likely events or consequences associated with the continued existence or demise of the Treaty.

If the ABM Treaty continues to exist, the next twelve years are likely to include the following:

- The United States and Russia will have negotiated amendments to the Treaty which permit three to five ABM fixed, ground-based launch sites limited to 100 missiles and launchers, no rapid reload launchers, and no multiple independently targeted warheads on the ABM interceptors. Such a deployment would permit an effective US NMD for limited ballistic missile threats without undermining Russia's deterrence capabilities, even at START II and START III force levels.
- START II will have been ratified by Russia's Duma, and START III will have been negotiated, ratified, and implemented by both parties. The result of these actions will be to further strengthen the Nuclear Non-Proliferation Treaty.
- British, French, and Chinese strategic ballistic missiles will be limited through an expanded multilateral strategic arms agreement ("START X"), or by mutual example.
- Space-based ABM and dual-capable ASAT systems will be banned, thus protecting space assets. This ban will also reinforce the premise of the Outer Space Treaty, which calls for freedom of exploration of outer space and celestial bodies.

- Space-based TMD systems will continue to be banned because of their inherent ABM capabilities.

If the ABM Treaty ceases to exist, the following events or consequences may take place between 1998 and 2010:

- Ballistic missile threats will increase, both regionally and intercontinentally in order to assure penetration of unlimited BMD systems.
- Rest of the World (ROW) satellites will be launched in increasing numbers. In addition to providing data supporting military operations, these satellites' launchers could readily be converted into ICBMs with unknown payloads.
- Chemical weapons and biological weapons on ballistic missiles may be used for purposes of intimidation or deterrence by ROW. ROW nations will also increase their dependency on cruise missiles in view of unlimited BMD systems. In response, the United States and its allies are likely to deploy highly-sophisticated defensive systems designed to counter both ballistic and cruise missiles. Escalating tensions will lead to serious and numerous regional and local conflicts, some of which will include the use of both ballistic missiles and TMD systems.
- Space-based lasers will be developed and deployed for purposes of both NMD and TMD.
- Ground- and space-based ASATs will be developed and deployed in order to achieve "space control."
- START II will remain stalled in the Russian Duma. There will be no START III. The Russians may withdraw from START I.
- There will be no "START X."
- A "Cold Peace" will prevail. Strategic and crisis stability will erode, if not be eliminated. The arms race will continue. The Russians will depend even further on ballistic missiles carrying MIRVed nuclear warheads in response to unlimited US ABM deployments.
- The Chinese will develop ABM and ASAT systems.

When the likely outcomes are weighed, it is clear that maintaining and strengthening the ABM Treaty is in the national security interests of Russia and the United States, and in the international security interest of all countries. Avoiding offense/defence arms races is in everyone's interests — politically, militarily, and economically.

Chapter Three

Outer Space: The Conference on Disarmament Dimension

Donald Sinclair*

In addressing the question of outer space from the Geneva or Conference on Disarmament (CD) perspective, it might be useful to address three questions:

- What are the prospects for work this year in the CD on outer space?
- What has the CD done in the past on this issue?
- Should an Ad Hoc Committee on Outer Space be established, what needs to be done?

The answer to the first question requires a bit of crystal-ball gazing. In February 1998, the CD agreed to its perennial eight-point agenda which includes the item "Prevention of an Arms Race in Outer Space." While the agenda was agreed to quickly this year, the CD has continued to struggle to establish an agreed work programme. For several delegations (not including Canada), priority work should focus on negotiating a transfer ban on antipersonnel landmines; for others, work should commence on nuclear disarmament. On the latter, there are a wide-range of options, including the unrealistic proposal from the Program of Action of the G-28 (the non-aligned among the 61 CD members) to negotiate the elimination of nuclear weapons in a time-bound framework — a non-starter for the P-5 and indeed all Western members of the Conference.

This year, Canada, South Africa, and Belgium have all made proposals on nuclear disarmament. Canada's proposal calls on the CD "to move now to proceed to substantive discussion of nuclear disarmament issues with a view to identifying one or more possible issues for multilateral negotiation." Canada added that it is not important what the mechanism is called; what is important is what it does. The South African proposal is a variant of this formula which draws in language from the NPT Principles and Objectives

---

* This chapter is an edited version of a presentation made to the Fifteenth Annual Ottawa NACD Verification Symposium, "Arms Control and the Rule of Law: A Framework for Peace and Security in Outer Space," 11–14 March 1998, Montebello, Quebec, Canada. The views presented in this chapter are those of the author and do not necessarily represent those of the Department of Foreign Affairs and International Trade or the Government of Canada.

Document relating to "systematic and progressive efforts to eliminate nuclear weapons." In addition, Canada and many others have called for negotiations to begin on a Fissile Material Cut-Off Treaty on the basis of the Shannon Report and the mandate contained therein, but this appears difficult under current circumstances.

It is necessary to have an understanding of the issues before the CD in order to understand why the Conference does not just get down to work on a single issue. Without agreement as to what that single issue should be, most delegations support a "balanced" work programme including both conventional and nuclear issues. Because this approach could lead to a "lowest common denominator" approach (NSAs and APLs), Canada and some others have called for a thorough airing in the Conference by a variety of mechanisms, including Presidential consultations and the appointment of facilitators and Special Coordinators to explore possible areas of agreement on all issues before the Conference.

It is within this complex web of issues, priorities, and, frankly, games being played that outer space takes its place. Last year, outer space fell off the CD priority list entirely. Many, including members of the P-5, argued that there was no arms race in outer space and therefore no need for any CD activity on this issue.

The situation has changed somewhat this year. In 1997, at the First Committee of United Nations General Assembly (UNGA), Sri Lanka modified its rather cumbersome text on "The Prevention of an Arms Race in Outer Space," in particular, the operative paragraph dealing with the CD. The 1997 text "invited" the CD to "re-examine the mandate contained in its decision of February 1992, with a view to updating it as appropriate, thus providing for the re-establishment of the Ad Hoc Committee during the 1998 session." This was a significant change in the CD context for a variety of reasons. The previous resolutions on the subject had "requested" the CD to re-establish the Ad Hoc Committee. This formulation ran afoul of the view of many in the CD that the CD is an independent body that does not take its directions or instructions from UNGA.[1] Secondly, it prejudged the mandate of the Ad Hoc Committee simply by ignoring the issue. The 1997 formulation avoided both these problems and Canada, for one, changed its vote to one in favour of the resolution. By loosening the language of their text, the Sri Lankans succeeded in putting pressure on those who are less keen to have any work on outer space done in the CD.

The receptivity of the CD this year to work on outer space appears to have changed as well. In its statement of 22 January 1998 Canada called on the CD to establish an Ad Hoc Committee on Outer Space with the mandate to negotiate a convention for the non-weaponization of outer space. Recognizing

1. One should note that this concept is often selectively applied.

that there are a variety of views on the proposal, Canada called for the appointment of a Special Coordinator with an appropriate mandate to explore prospects for the early establishment of such an Ad Hoc Committee.

To date, the response has been cautious but generally favourable — at least by those who have spoken in Plenary to date. India, for example, welcomed the Canadian proposal and called for the CD to address urgently a ban on ASAT weapons and to develop "rules of the road" arrangements in space in order to ensure its peaceful use. Austria, The Netherlands, and Germany also supported the Canadian approach.

Of the P-5, the Russian Federation supported either the establishment of an Ad Hoc Committee on Outer Space with the mandate of previous years or the appointment of a Special Coordinator to examine possibilities for a new mandate. Ambassador Berdennikov added that "in principle, we perceive with interest the Canadian ideas concerning non-weaponization of outer space." France identified the prevention of an arms race in outer space as one of its three priority issues for the CD (including antipersonnel landmines and a fissile material cut-off treaty). The Ambassador noted that France shared most of the views expressed by Canada and supported the efforts of Sri Lanka and Egypt to launch negotiations on this issue. She recalled the changes made last year to the annual UNGA resolution on outer space which allowed it to pass without a negative vote.

Despite this show of support for work on the non-weaponization of space, the United States has yet to give its formal views in the CD Plenary on the subject. In the past, the US has opposed an Ad Hoc Committee on Outer Space in the CD with a negotiating mandate, and in recent years, has been reluctant to support the establishment of an Ad Hoc Committee with the traditional mandate to explore relevant issues. This was less of an issue in the years 1993–1995 when the Comprehensive Nuclear Test Ban Treaty (CTBT) negotiation occupied the time of the CD. For its part, China, at the 51st session of UNGA last fall, called for the prohibition of the development of outer space weapons.

The current Swiss President of the CD is negotiating language which, if accepted, could lead to the appointment of a Special Coordinator on this issue. The US has indicated a willingness this year to support the appointment of a Special Coordinator, depending on the wording of the mandate, which now seems close to agreement. The best we can realistically hope for from this person's work will be an agreement to establish an Ad Hoc Committee with a mandate similar to previous ones — that is, without a negotiating mandate but with the obligation to explore all relevant issues and proposals. This would be a valuable step forward.

This brings us to the second question: what has the CD done in the past on outer space?[2] The CD established an Ad Hoc Committee on Outer Space

2. Here I am relying on records of the Conference rather than first-hand experience.

from 1985 to 1992 with the mandate each year to "examine, and to identify, through substantive and general considerations, issues relevant to the prevention of an arms race in outer space." In so doing, the Committee "will take into account all existing agreements, existing proposals and future initiatives as well as developments which have taken place since the establishment of the Ad Hoc Committee in 1985."

A quick glance through the Final Records and CD Working Papers of this period reveals the wealth of useful material that already exists. It also reveals the limits to which an Ad Hoc Committee without a negotiating mandate can go. In the early days, the initial tasks were to explore existing legal instruments to identify gaps in the legal regimes governing outer space and to try to develop common understandings of definitions and concepts. The United Kingdom, Canada, the "Group of Socialist Countries," and others were active in this regard. In 1986, and expanding on this in 1988, Canada produced papers on the use of certain terms relating to arms control and outer space in an effort to clarify the issues under debate. The 1988 paper focussed on the term "weaponization" and its meaning, pointing to the clear difference between weapons or "devices or installations capable of attacking, damaging or disrupting the functioning of spacecraft in space, or of objects in air, on land or at sea," and military satellites for command, control, and communications.

Sweden called for a Treaty to ban all space weapons, including weapons directed against targets in space. They also called for the prevention of "the harmful interference in the normal functioning of permitted space objects" and for the multilateralization of the bilateral US-USSR 1972 ABM Treaty ban of "the development, testing and deployment of space-based ABM systems."

Venezuela, supported by other Latin American states, proposed an amendment to the Outer Space Treaty to add the phrase "any kind of space weapon or system of space weapons" after the injunction against the use of weapons of mass destruction (WMD). Australia and Canada proposed a strengthening of the Convention of the Registration of Objects Launched into Outer Space. Others proposed mechanisms designed to protect objects in space or to establish clearer rules for the peaceful use of space — the so-called "rules of the road" approach recently referred to by the Indian Ambassador to the CD.

France elaborated on its proposal for an international satellite monitoring agency (ISMA), first proposed at UNSSOD I, by advocating as a first phase of the ISMA the establishment of an agency for the processing of satellite images. France argued that this agency would be a confidence building device and would not be intended to be the embryo of a verification system with universal competence attached to the UN. A Soviet proposal at UNSSOD III in 1988 went further in declaring support for an ISMA, based on French and

Canadian ideas, "which would provide the international community with information relating to compliance with multilateral arrangements in the field of disarmament and the reduction of international tension ..."

Clearly, any Ad Hoc Committee on Outer Space, should it be established in 1998, should take account of and update previous work. The early identification of the current position of states, including the many new members of the CD will be important. Intervening events will also have an impact. For example, of the P-5, China was alone during the CTBT negotiations in arguing for a space-based verification role.

Where Do We Go From Here?

In a 1988 address to the CD, the then-Canadian Ambassador to the CD, de Montigny Marchand, outlined four considerations for CD work on outer space:

- work must enhance stability and not detract from it;
- CD negotiations must complement negotiations between the two major space powers;
- the CD must recognize that a very considerable measure of prohibition and protection already exists in outer space and must base its efforts on that foundation; and,
- the CD must not lose sight of the very practical division of labour between the CD and the United Nations Committee on the Peaceful Uses of Outer Space.

This was a good beginning, perhaps, in 1988, reflecting a Cold War perspective. But, what is to be done in 1998? First, Canada must continue its work, reflection, and appropriate studies on outer space issues on the assumption of the early start to work in Geneva. Second, it should review existing work, in particular with respect to existing legal instruments pertaining to outer space and the identification of gaps, as well as review and update the work on concepts and definitions already undertaken. The careful examination of legal issues pertaining to outer space (including possibilities for reinforcing the Outer Space Treaty) will help define the scope of any future negotiated agreement. Third, work should progress on identifying and giving content to elements of a Convention for the Non-Weaponization of Outer Space. Measures relating to scope, confidence building, transparency, verification, and compliance will be especially important, but work could begin now. Fourth, and related to the scope of any non-weaponization convention, will be the need to assess the impact of BMD on outer space arms control measures. Finally, the Canadian delegation in Geneva must work to secure agreement in the CD on a Special Coordinator with an appropriate mandate to explore possibilities for work in the CD and then proceed to establish an Ad Hoc Committee with an appropriate mandate.

Chapter Four

Banning the Stationing of Weapons in Space Through Arms Control: A Major Step in the Promotion of Strategic Stability in the 21st Century

F.R. (Ron) Cleminson*

Arms control could prove useful in constraining the fielding and employment of certain capabilities which in the hands of potential adversaries might, on balance, be more harmful to us than like technologies in our hands might benefit us.[1]

General John M. Shalikashvili, USA (ret'd)
9 June 1998

Setting The Stage

These days, credible global threats to US national security are more difficult to come by. With the Cold War behind us and the Revolution in Military Affairs (RMA) ahead, American strategic policy planners are engrossed in trying to identify the means by which US national security objectives can be met and peace and security on an international level reinforced. Clearly, the United States shoulders a leadership role in the search for viable options. In the past, planners for national security purposes have had the tendency to think mostly in terms of the acquisition of "more and better" weapons systems when attempting to offset perceived threats. The theoretical "weapon of choice" has been the sustained deterrent effect of military power projection. Today, however, cost is becoming an increasingly important factor in the security equation. As the *Christian Science Monitor* observed only recently "with US defense spending constrained (sic) for the foreseeable future to

---

* This chapter is an edited version of a presentation made to the Fifteenth Annual Ottawa NACD Verification Symposium, "Arms Control and the Rule of Law: A Framework for Peace and Security in Outer Space," 11–14 March 1998, Montebello, Quebec, Canada. The opinions expressed in this chapter are those of the author and do not necessarily reflect those of the Department of Foreign Affairs and International Trade or the Government of Canada.

1. General John M. Shalikashvili (ret'd), keynote address at the DWSA 7th Annual Conference on Controlling Arms, Philadelphia, 9 June 1998.

about $250 billion per year, ... the need for less esoteric hardware will be more critical to preserving US military power."[2] Cost-effective alternatives, of which arms control is one, should logically receive greater attention.

Until now, the application of arms control measures to the arena of national security policy has seldom been considered as a serious and feasible option for the policy-planning elite. This is despite the fact that, for more than 30 years, NATO promoted an alliance security policy of "undiminished security at lower levels of armament." Those lower levels of armaments were achieved finally through the arms control and disarmament process mandated by the United States and its allies in the multilateral CFE Treaty. Nevertheless, arms control as a means of contributing significantly to the structures of national security has remained a concept which has been very much underplayed. In the decade to come, NATO and its members might usefully redefine their respective security policies for the 21st Century as "undiminished security at lower levels of defence spending." Nowhere will this have greater resonance than in the international community's approach to the management and use of outer space.

The 1972 Anti-Ballistic Missile (ABM) Treaty is the most graphic illustration of the application of arms control for the purpose of constraining deployment of weapons systems. It has served as a vibrant example of what General Shalikashvili characterized as a means of "constraining the fielding and employment of certain capabilities." The ABM treaty stands out today not only as a proven model for the non-proliferation of weapons but also for the "non-expenditure of funds." One can only imagine what the world might have looked like had the ABM Treaty, which constrained development and deployment of ABM systems, not come into existence.

In Washington D.C., on 5 May 1998, Sandy Burger, President Clinton's National Security Adviser, told the annual meeting of Business Executives for National Security that "the future of arms control is now." Mr. Burger was referring, in part, to changes in the ABM Treaty which are up for approval in the United States Senate and the Russian Duma. Though bilateral in nature, the ABM Treaty has become multilateral in scope in that it is now recognized internationally as part of a body of treaties and agreements which constitutes "customary space law."

The lessons which have been learned within the confines of the ABM Treaty experience have a much broader potential application in terms of the outer space issue. Back in 1991, according to a report in *Aviation Week and Space Technology*, the real question related to national security and the management of outer space was framed as "whether to ban weapons in space."[3] Here, the concept of using arms control as a mechanism for "constraining the

---

2. "Global Report," *The Christian Science Monitor*, 19–25 December 1997.
3. Bringing SDI Down to Earth," *Aviation Week and Space Technology* (18 February 1991), 7.

fielding and employment of certain capabilities" has explicit application. It is more relevant today than it has been in the past and will continue to be so in the future — both in terms of protecting assets in space and in preventing the placement of weapons systems in space. While the tactical threat faced in the Gulf War has served to underscore and validate the requirement for a theatre missile defence, the extraterrestrial threat which might warrant the fielding of an extensive BMD capability remains unacknowledged. At the same time, the increasing commercialization of space makes the management of that environment in its totality a matter of international concern. The international community, therefore, has a right and a responsibility to become engaged in initially analyzing and eventually managing the interrelationship between the commercial exploitation of space and the legitimate concerns related to peace and security in outer space.

Overview

The dramatic improvement in East-West relations over the last decade has brought about a major change in the international security environment. This has been dramatized in a series of significant steps in the process of multilateral non-proliferation, arms control, and disarmament (NACD) affairs. These steps included the sequential conclusions of the 1993 Chemical Weapons Convention (CWC) and the 1996 Comprehensive Test Ban Treaty (CTBT) and the successful 1995 review of the Non-Proliferation Treaty (NPT). Additionally, the effective implementation of the bilateral Intermediate-Range and Shorter Range Missiles Treaty (INF) by the United States and the Russian Federation, in which an entire class of nuclear weapons systems was eliminated, is another vivid and practical illustration of arms control at work.

In the political arena, with obvious exceptions, there continues to be a dominant trend toward the replacement of confrontation with dialogue and of rivalry with cooperation. In addition to the traditional NACD issues, the world as a whole is increasingly faced with non-military threats to international security by non-traditional security challenges (NTSCs) such as cross-boundary environmental pollution, ecological degradation, off-shore harvesting of resources, and in the global commercialization of the exploitation of space. It is in this latter category that multilateral discussion is required to ensure that space remains free of weapons systems, whether they be conventional or weapons of mass destruction (WMD).

Space Technology

Technology by itself does not threaten anyone. It should not be seen as an inhibitor to multilateral discussions in this area. Efforts to steer the direction of technological change can only make progress if the realities of contemporary life are taken into account. The current state of technological advancement represents decades of cumulative knowledge that cannot be erased. It is also unrealistic to believe that the process of technological innovation can be frozen to prevent its military application. However, the very technologies capable of improving existing weapon systems and developing new ones can also be used to constrain the fielding and employment of such systems, particularly in terms of their use in space. A few of the many areas in which militarily capable technologies could actually at the same time promote rather than threaten international security are the use of communication technologies for advance warning of impending conflicts, the employment of remote-sensing techniques for NACD verification, and the development of appropriate techniques for monitoring and verification.

Space technology, like space law, is not a field unto itself. Drawing upon a multitude of different scientific disciplines, space technology could be seen as a collection of new technologies directed toward the exploration and exploitation of the vast region of the universe beyond the Earth's atmosphere. Practically every area of modern science has fed space technology. For the first time, the proliferation of these technologies is being led by the commercial sector, with more than $120 billion of space assets likely to be placed in orbit in the next five years by the United States alone.[4]

Since the first satellite was placed in orbit in 1957, space capabilities have developed essentially in four areas: space transportation, sensors, spacecraft, and ground segment. Military activities in space have been devoted toward five traditional support missions: communications, reconnaissance and surveillance, navigation, meteorology, and geodesy. The increasing commercialization of space, projected to top $500 billion over the next decade, will take advantage of these technologies in what appears to be an almost unlimited number of applications ranging from farm management to disaster relief.

Anti-Satellite and Satellite Defence Weapons

Not unrelated to the concept of a ban on weapons in space, and, in fact, emerging in its absence, is the perceived need to develop an ability to destroy "enemy" satellites and to defend "friendly" ones. This category comprises all the features of a military competition and weapons proliferation in space: (a) mines, directed-energy weapons, kinetic energy weapons, jammers, and electronic countermeasure pods to destroy or fool enemy satellites; (b) defensive escorts for friendly satellites, carrying jammers, decoys, shields, or weapons to fight off anti-satellite weapons (ASATs); and, (c) spacetracking and identification sensors for such missions as well as treaty monitoring.

Although military support missions have been routinely carried out and are considered to be stabilizing, space has not yet become a venue for weapons deployment. As of now, there are in excess of 8,000 human-made objects in space. Not all of them are equally capable of performing military support

4. "Global Report," *The Christian Science Monitor Weekly*, 19–25 December 1997.

missions. There are also vast differences in the degree of their technological sophistication. Considerable amounts of human, technical, and financial resources are known to have been allotted for safeguarding existing space-based assets, protecting them from destruction by an adversary, and acquiring the technological means to destroy space-based systems.

Beyond the esoteric discussions centering on the feasibility or advisability of developing an ASAT capability is a more subtle but nonetheless telling conundrum. With the proliferation of satellites for increasingly diverse purposes deployed by a wide range of sponsors, what is an "enemy" satellite? During the 1992 Gulf War, for example, both Coalition and Iraqi forces used channels on ARABSAT. Had that satellite had been destroyed, which side would have benefitted?

The Existing Legal Regime

It is sometimes contended, though not with great credibility or conviction, that the existing legal regime in space is adequate for today and, more importantly, for tomorrow.[5] In fact, space law originated in the 1950s and was based on extractions from certain international conventions including many which originated in the United Nations-sponsored efforts and forums. It is not a new, independent branch of law custom-made for space, but rather has developed over time and out of existing agreements. Thus, it should be recognized that today there is no single treaty which fully specifies what activities in outer space are permitted or prohibited.

Jurists of diverse nationalities point to the Charter of the United Nations, the Limited as well as the Comprehensive Test Ban Treaties, the Outer Space Treaty, the SALT I and II and START I and II Agreements, and the ABM Treaty and Protocol to the ABM Treaty as the core body of work that defines the arms control and disarmament dimensions of outer space. A host of other non-proliferation, arms control, and disarmament treaties, agreements, and resolutions make significant contributions to the current legal regime of outer space through direct reference to its domain (e.g., Environmental Modification Convention), or through establishing legal precedents or international norms (e.g., The Antarctic Treaty).

Although more than a dozen international agreements, including the UN Charter, make reference to outer space, there are currently four *major* treaties in force that are viewed as being most significant in the regulation of various aspects of the use of space for military purposes:

- The 1963 Treaty Banning Nuclear Weapon Tests in the Atmosphere, in Outer Space and Under Water (the "Partial Test Ban Treaty") prohibits nuclear explosions such as nuclear-weapons tests in space.

---

5. CD/PV.775, 21 August 1997, 9.

- The 1967 Treaty on Principles Governing the Activities of States in the Exploration and Use of Outer Space, including the Moon and Other Celestial Bodies (the "Outer Space Treaty") bans the stationing of weapons of mass destruction in space. Often termed "the cornerstone of the outer space region," it also bans weapons testing and military manoeuvre on celestial bodies.
- The 1972 Treaty between the United States of America and the Union of Soviet Socialist Republics on the Limitation of Anti-Ballistic Missile Systems (the "ABM Treaty") prohibits the testing, development, and deployment of space-based ABM systems or components.
- The 1996 Comprehensive Test Ban Treaty (CTBT) bans nuclear weapons testing in all environments.

Not all of the existing agreements relating to outer space necessarily apply with uniformity to all countries. The mosaic of international law, said to guide and govern approaches to and activities in space, consists primarily of a variety of multilateral agreements, including the UN Charter itself. Thus, it can be argued that the legal regime is in need of regularization and that it should be brought up-to-date to take into consideration the burgeoning of new technologies, the mushrooming of commercial use, and the new peace and security structure occasioned by new NACD treaties, both bilateral and multilateral.

Confidence Building in Space[6]

In addition to negotiated treaties designed to establish a level playing field in outer space, there is a need to foster confidence in a less structured manner by enhancing transparency. The removal of suspicion about intentions is thus a general interest, even for military powers with known space-based military capabilities. Several new technologies have already made it possible to increase transparency by the use of remote sensors for promoting confidence among space-based military powers and for facilitating the conclusion of several types of arms control agreements. The central benefit of these types of systems is that, theoretically at least, they would permit verification without compromising the integrity of weapons systems and jeopardizing military security.

Much of the debate and negotiation of issues relating to outer space has occurred and will continue to occur in international bodies such as the Conference on Disarmament (CD) and the Committee on Peaceful Use of Outer Space (COPUOS). Although the mandates of CD and of COPUOS are distinct, they are increasingly being seen as mutually reinforcing. This is an area in

---

6. For a comprehensive review of the subject see "Study on the Application of Confidence-Building Measures in Outer Space," Study no. 27 (New York: Centre for Disarmament Affairs, United Nations, 1994).

which continued and increased cooperation is not only desirable but is, indeed, necessary. It would be useful, therefore, for the CD to focus in some detail on the peace and security dimensions of the evolving and increasing commercial use of space. The significant expansion of commercialization in space, without an adequate regulatory environment (i.e., no effective regulatory regime to guide and adjudicate) could lead to misunderstanding and conflict. The CD shares the responsibility to ensure that this does not occur. Understanding outer space, as it exists today and as it is evolving, is part of the CD's mandate in this regard.

Evaluation of the Adequacy of the Existing Regime

A perusal of international law as it exists at present reveals that in reality very few military uses or activities in outer space are actually prohibited. An analysis of current international agreements covering military space activities indicates that, while some military space options are prohibited, particularly under the 1967 Outer Space Treaty and the 1972 ABM Treaty, there are many loopholes and lacunae and many more provisions open to a variety of interpretations. The goals of ensuring that outer space remains as a "weapons-free" environment or, put another way, that space remains "non-weaponized" are still unfulfilled.

It should be clearly understood that, other than weapons of mass destruction, the current space law regime does not preclude the stationing or use of weapons in outer space. Hence conventional and/or "exotic" weapons are, arguably, permissible in space. One might ask rhetorically whether this is what the international community wants.

We must recognize that the situation in which we find ourselves at this time is not conducive to curbing the "migration" of terrestrial-based weapons into space. Much needs to be done, therefore, to improve and update the current state of international agreements and to discourage such a migration with its attendant consequences for international peace and security. Therefore, it is difficult to imagine that, as some continue to claim, the existing legal framework is adequate to meet the security concerns of the next century.

Areas for Consideration by the CD in 1998 and 1999

As the CD assesses its mandate related to peace and security in outer space, there are numerous questions which could be usefully addressed during the 1998/99 sessions. While the list of questions below clearly is not exhaustive, it does identify points of focus and issues of immediate concern which could be discussed:

- How can international legal instruments applicable to outer space be modified or extended through protocols to incorporate advances reached in other areas of arms control and disarmament?

- How can we take advantage in the area of outer space of the growing acceptance elsewhere of multilateral solutions through arms control?
- How can the emerging array of confidence building measures in other areas best be applied to enhance security in outer space?
- How can we create an arms control regime in outer space which makes allowance for the dual nature of many technologies?
- How can the existing regime to promote the peaceful uses of outer space be used to assist non-proliferation?
- How can the work of COPUOS and the CD be better coordinated, particularly in terms of "Outer Space and the Existing Legal Framework" while maintaining and respecting their separate mandates?

Conclusions

- Today, no single treaty fully specifies which activities in outer space are permitted and which activities in outer space are prohibited. Furthermore, all existing agreements relating to outer space do not necessarily apply uniformly to all countries. The international framework of customary law said to guide and govern approaches to and activities in space consists primarily of a variety of multilateral agreements, including the United Nations Charter itself.
- Space law originated in the 1950s and was based on conventions adopted by the United Nations. It is not a new, independent branch of law custom-made for space, but rather, it has developed over time and out of existing agreements.
- Although the basic principles relating to the maintenance of peace and security in outer space have not changed appreciably since the inception of human use of outer space, the dynamics of the use of space continue to evolve and at an increasing rate. In the first half of this decade, the CD's attention in accordance with its mandate was focussed elsewhere. Now, with the successful negotiation of the CWC and the CTBT behind it, the CD should re-position itself to deal with and focus on the multilateral aspects of outer space.
- With scientists and engineers of many nations preparing to launch the International Space Station on the one hand, and with very clearly enunciated concerns about trends pointing in the direction of the weaponization of space on the other hand, the time is opportune for the international community in its deliberative and negotiating bodies to examine the outer space issue in all its aspects and to take appropriate action.
- The venue within which much of the debate and negotiation of issues relating to outer space has and will continue to occur includes other bodies such as COPUOS. Although the mandates of COPUOS and the CD are distinct, they are increasingly seen as mutually reinforcing. This is an area

in which continued and increased cooperation is not only desirable but is, indeed, necessary.

- It would be useful, therefore, for the CD to focus, in some detail, on the peace and security dimension of the evolving commercial use of space and of its relevance to activity related to international security issues, including non-proliferation of weapons and arms control.
- The significant expansion of commercialization in space without an adequate regulatory environment (i.e., no central authority to guide and adjudicate) could lead to misunderstanding and conflict in a non-traditional manner. The CD shares the responsibility to ensure that this does not occur. Understanding outer space as it exists today and as it is evolving is part of the CD's mandate in this regard.

Chapter Five

Recent Developments in Space Law

M. Lucy Stojak*

Introduction

In July 1985, Canada submitted a substantive Working Paper to the Conference on Disarmament (CD) entitled *Survey of International Law Relevant to Arms Control and Outer Space*. As stated in its preface, the working paper was meant to provide a review of international law relating to arms control and outer space. Its objective was to provide a rational basis for discussion from which the CD *ad hoc* Committee for the Prevention of an Arms Race in Outer Space might wish to develop its approach to the subject.

The working paper indicated that the following military activities in space were consistent with existing international law:

- the use of military personnel;
- the use of space-based remote sensors for military purposes; and,
- the use of space-based communication, navigation, and meteorological systems for military purposes.

Certain other military activities in space were clearly inconsistent with international law. These comprised:

- interference with space-based remote sensors used for military purposes as between the US and former USSR;
- placement of nuclear weapons and other weapons of mass destruction in orbit around the earth and on celestial bodies or in orbit around them;
- placement of military bases and conduct of military tests or manoeuvres on celestial bodies and in orbits around them;
- testing of nuclear weapons in outer space;
- development, testing, and deployment of space-based anti-ballistic missile systems or components; and,

---

* This chapter is an edited version of a paper commissioned for the Fifteenth Annual Ottawa NACD Verification Symposium, "Arms Control and the Rule of Law: A Framework for Peace and Security in Outer Space," 11–14 March 1998, Montebello, Quebec, Canada.

- military or hostile use of environmental modification techniques in outer space.

Much of the debate surrounding permitted and prohibited activities focused on certain key definitions such as "peaceful purposes," "free use," "militarization," and "weaponization."

Thirteen years later, little if any progress has been made on this front. Though extensive state practice certainly supports the interpretation of "peaceful purposes" as meaning "non-aggressive purposes," many states still do not officially recognize and accept this interpretation. The Missile Technology Control Regime (MTCR) guidelines discussed later in this chapter are criticized by some as being in contravention of the freedom of use principle applicable to outer space. Some states persist in demanding the "demilitarization" of outer space, an unrealistic demand in view of the stabilizing role played by certain military satellites and their value to ensuring and maintaining peace.

In researching this chapter, it became clear that three of the most frequently recurring concepts concerning outer space were: dual-use, control (effective control), and access to outer space. This chapter focuses on the following four main topics in terms of the existing international law: Global Positioning Systems (GPS), remote sensing, the International Space Station, and the MTCR guidelines. All of these reflect issues raised by either dual-use technologies, controlling outer space, and access to outer space. The final section contains a summary of proposals made to amend the Outer Space Treaty and assesses the likelihood of such amendments actually being made. The chapter concludes with a series of modest recommendations to reactivate interest in the CD on the topic of the prevention of an arms race in outer space.

Global Positioning Systems

Applicable International Law

Of importance here to the discussion of state responsibility for providing satellite navigation services are the Outer Space Treaty[1] and the Liability Convention,[2] both of which the United States is a party to. This is a relevant point to make since, at present, only the US has an operational GPS.

For purposes of this discussion, the relevant liability provisions are found in Articles VI and VII of the Outer Space Treaty, which provide:

---

1. Treaty on Principles Governing the Activities of States in the Exploration and Use of Outer Space, Including the Moon and Other Celestial Bodies, 18 U.S.T. 2410, T.I.A.S. No.6347; 610 U.N.T.S. 205, entered into force 10 October 1967 (hereinafter Outer Space Treaty).
2. Convention on International Liability for Damage Caused by Space Objects, 24 U.S.T. 2389, T.I.A.S. No. 7762, 961 U.N.T.S. 187 entered into force 1 September 1972 (hereinafter the Liability Convention).

> States Parties to the Treaty shall bear international responsibility for national activities in outer space, including the moon and other celestial bodies, whether such activities are carried on by government agencies or by non-governmental entities, and for assuring that national activities are carried out in conformity with the provisions set forth in the present Treaty. The activities of non-governmental entities in outer space, including the moon and other celestial bodies, shall require authorization and continuing supervision by the appropriate State Party to the Treaty. When activities are carried on in outer space, including the moon and other celestial bodies, by an international organization, responsibility for compliance with this Treaty shall be borne both by the international organization and by States Parties to the Treaty participating in such organization.[3]

> Each State Party to the Treaty that launches or procures the launching of an object into outer space, including the moon and other celestial bodies, and each State Party from whose territory or facility an object is launched, is internationally liable for damage to another State Party to the Treaty or to its natural or juridical persons by such object or its component parts on Earth, in air space or in outer space, including the moon and other celestial bodies.[4]

The Liability Convention elaborated on the liability provisions of the Outer Space Treaty. Articles II and III provide:

> A launching State shall be absolutely liable to pay compensation for damage caused by its space object on the surface of the earth or to aircraft in flight.[5]

> In the event of damage being caused elsewhere than on the surface of the earth to a space object of one launching State or to persons or property on board such a space object by a space object of another launching State, the latter shall be liable only if the damage is due to its fault or the fault of persons for whom it is responsible.[6]

Thus, the regime of liability under the Liability Convention will depend on where the space object inflicts the damage. If on the ground or to aircraft in flight in the air space, then the launching state will be held absolutely liable. If in outer space or to another spacecraft in flight, then liability will exist if the launching state was at fault.

While the above cited provisions seem clear in creating a regime of liability based on the *situs* of the damage, the convention is not clear on the type of damages recoverable. Article I defines damage as:

---

3. Outer Space Treaty, supra note 1, Article VI.
4. *Idem*, Article VII.
5. Liability Convention, supra note 2, Article II.
6. *Idem*, Article III.

> ... loss of life, personal injury or other impairment of health; or loss of or damage to property of States or persons, natural or juridical, or property of international government organizations.[7]

Reading Articles I, II, and III together, it is clear that these damages are recoverable if they directly result from the physical impact of a space object. What is unclear is whether damages that allegedly flow from a space object absent any direct or physical causation are recoverable. Stated another way, was the Liability Convention meant to cover consequential, economic, or indirect damages arising from the use of navigational satellite services provided by a state party to the agreement?

With regard to GPS-provided services, the United States certainly can be considered a "launching State,"[8] and the GPS satellites in the constellation "space objects,"[9] according to the definitions found in Article I of the Liability Convention. As such, if when launching a space object, say a replacement GPS Block II R satellite, direct damages are caused, the launching state would be liable.

On the other hand, indirect damages must be distinguished. The Convention is unlikely to apply in cases where damage arises not directly because of the physical impact of a space object, but indirectly because a space object such as a GPS satellite may have transmitted an erroneous or improper navigation signal causing an aircraft to crash. Neither the language of the Convention, the negotiations leading to its passage, nor state practice would support a claim for damages sustained in the context of an alleged negligently provided GPS signal. While the pre-treaty deliberations did raise the issue of indirect damages, the discussions seemed to focus on the consequential damage resulting after the crash of a space object. One commentator has stated that it is doubtful these deliberations fathomed damage caused by a space object still functioning in space.[10]

Other evidence supports this position. First, in Congressional documents prepared for ratification hearings before the US Senate, indirect damages were raised in the context of electronic interference from an orbiting satellite. The Senate indicated that the liability for space activities did not include recovery for non-physical damages, and that the US position before the United Nations, stated as early as 1971, was that indirect damages were not covered under the

7. *Idem*, Article I(a).

8. Article I(c) defines launching State as "(i) a State which launches or procures the launching of a space object" or "(ii) a State from whose territory or facility a space object is launched."

9. Article I(d) defines space object to include "component parts of a space object as well as its launch vehicle and parts thereof."

10. C.Q. Christol, *The Modern International Law of Outer Space* (New York: Pergamon Press, 1982), 95–100.

Liability Convention.[11] More recently, one author has expressed doubts that the Liability Convention would apply to similar cases of economic damages caused by direct broadcast or remote sensing satellites.[12]

In short, it is unlikely that the United States would recognize the validity of a claim for indirect damages arising out of GPS-provided services under existing Treaty law. The issue of indirect damages has never been resolved on the international level,[13] and the United States has consistently taken the position that indirect damages are not cognizable under the Liability Convention.

Policy Concerns

GPS reduces the reliance on terrestrial navigation systems and provides precise information for navigation, targeting, and troop movement. The Gulf War demonstrated the value not only of military uses of GPS but of civilian uses such as mapping, transportation, and air traffic. GPS is thus a prime example of a dual-use technology which raises conflicting questions of national security and economic goals.

In January 1993, the US Departments of Defense and Transportation (DoD and DoT) formed a joint task force to study the management and operation of GPS as a dual-use (military and civilian) system. Key questions raised dealt with:

- who should manage the system?;
- how should it be funded?;
- the accuracy of the SPS signal;
- user concerns of continued integrity and availability of the signal;
- regulation of GPS augmentation system;
- international acceptance of GPS as a global standard; and,
- spoofing and jamming risks to the signal.[14]

---

11. K.K. Spradling, "The International Liability Ramifications of the US NAVSTAR Global Positioning System," (1990) *Proceedings of the Thirty-Third Colloquium on the Law of Outer Space*, 93, at 97.
12. B.A. Hurwitz, *State Liability for Outer Space Activities in Accordance with the 1972 Convention on International Liability for Damage Caused by Space Objects* (The Netherlands: Kluwer Academic Publishers, 1992), 253–255.
13. Some argue that claimants should be able to proceed under the convention for indirect damages arising out of the use of satellite navigation services if they can show causation. P.B. Larsen, "Legal Liability for Global Navigation Satellite Systems," (1993) *Proceedings of the Thirty-Sixth Colloquium on the Law of Outer Space*, 69, at 70. See also, W.F. Foster, "Convention on the International Liability for Damage Caused by Space Objects," (1972) *The Canadian Yearbook of International Law*, 137.
14. Joint Task Force Report quoted in Larsen, *supra*. For additional information see also, J.A. Rockwell, *Liability of the United States Arising out of the Civilian Use of Global Positioning System*, unpublished Master of Law Thesis, McGill University, August 1996.

While the task force report identified several issues requiring further study, it made at least three important recommendations:

- that a joint executive board comprised of members of the DoD and DoT Positioning/Navigation Executive Committees be established to resolve management and policy issues by consensus;
- that the DoD continue funding the existing system and the DoT fund any augmenting systems; and,
- that a study of all differential GPS augmentations be conducted to meet the needs of civil navigation applications.

The US Congress, as part of the National Defense Authorization Act for fiscal year 1994, requested a study of the future funding and management options for GPS. The study, conducted by the National Academy of Public Administration (NAPA), found that most aspects of GPS funding and management were sound. NAPA recommended that DoD continue funding and operating the system, but recognized the public service and safety applications to other civil agencies such as the DoT. They also recommended that Selective Availability, the technique employed by DoD to degrade the GPS signal, be turned off based on its added costs to the commercial market and its effect on international confidence in relying on GPS as the global standard.[15]

In January 1996, RAND Corporation's Critical Technologies Institute released its report, which was commissioned by the White House Office of Science and Technology. The RAND study highlighted the need for a clear policy directive on GPS, the lack of which has contributed to the ongoing debate on GPS-related issues such as the balancing of national security and economic interests, the funding and governance of the system, and the concerns of foreign users. Specifically, in addition to calling for the issuance of a statement of national policy, the RAND study concluded that the system should be operated and funded solely by the US government.[16] Direct user fees would not be charged for several reasons, namely, to discourage the development of foreign competing systems, to foster the adoption of the US system as a global standard, and to maintain the system under US control

---

15. See National Academy of Public Administration, National Research Council, *The Global Positioning System — Charting the Future, Summary Report* (Washington, D.C., May 1995).

16. The RAND study identified six different GPS institutional options as: (1) the present US military system; (2) a US civilian agency; (3) a privatized US entity; (4) a privatized international entity; (5) an augmented system made up of civil, private, and foreign elements; and, (6) a private system totally displacing GPS. There were five institutional conditions identified as necessary for continued operation of GPS: (1) frequency allocations from the International Telecommunications Union; (2) continued ground station sites in the US, UK, and Marshall Islands; (3) skilled operators; (4) funding; and, (5) procedural disciplines, especially in times of crisis for security reasons. Only the military institutional option could meet all five conditions at present.

rather than under the control of a potential adversary or international civil organization. The study recommended that the US engage in discussions with Japan and Europe on regional security and economic issues with an eye toward reaching international agreements.

In March 1996, the White House issued a formal policy in the form of a Presidential Directive concerning the civil and commercial use of GPS.[17] The directive sets policy goals and guidelines in addition to setting out the roles and responsibilities of involved US government agencies. The policy goals seek to: (1) strengthen and maintain national security; (2) encourage acceptance and integration of GPS into peaceful civil, commercial, and scientific applications worldwide;[18] (3) encourage private sector investment in, and use of, US GPS technologies and services; (4) promote safety and efficiency in transportation and other fields; (5) promote international cooperation in using GPS for peaceful purposes; and, (6) advance US scientific and technical capabilities.

Based on these guidelines, the DoD will continue to be responsible for maintaining and operating the basic GPS and providing the SPS signal on a continuous, worldwide basis. In addition, the DoD will continue to assess the national security implications of the civil use of GPS and develop measures to prevent its hostile use without unduly disrupting or degrading civil uses. The DoT will serve as lead agency for all federal civil GPS matters and continue to develop government augmentations for transportation applications. The DoT will also promote the acceptance of GPS as the standard in the national and international transportation systems. The State Department will maintain overall responsibility for developing bilateral and multilateral guidelines and agreements on the use of GPS.

Future International Framework

No single international organization exists to address the wide range of issues associated with the international use of GPS. Competing domestic interests concerning national security and economic issues within the US are only amplified when they become international security and economic concerns in international fora. While an organization like the International Civil Aviation Organization (ICAO) has the necessary expertise in technical matters, such as the integration of aviation safety standards worldwide, it is asking too much of the organization to try to balance national and international security concerns, along with states' highly politicized commercial concerns, to create a workable multilateral legal framework.

---

17. The White House, Office of Science and Technology Policy, National Security Council, "US Global Positioning System Policy," 29 March 1996.

18. Prior to the 1996 directive, the US made written offers to the International Civil Aviation Organization (ICAO) to make GPS available on a continuous, worldwide basis, and free of direct user fees.

There already exist at least fifteen international agreements relating or referring to GPS, and there are no legal barriers preventing states from entering into agreements with the United States to use GPS services.[19] Thus, given the absence of any legal barriers, bilateral agreements will likely be the future of international law with regard to GPS-provided services.

International Space Station

When the Outer Space Treaty and the various UN Resolutions adopted prior to the drafting and ratification of the Outer Space Treaty were adopted, both superpowers had launched numerous military satellites since 1957. The comment often made is that first, more restrictive language than that found in Article IV of the Outer Space Treaty could easily have been adopted and second, nations could have officially voiced their objections to certain types of military activities. It should be recalled that no country has formally objected to the use of outer space for military purposes.

Article 31(3) of the Vienna Convention on the Law of Treaties[20] provides that in the process of interpretation "any subsequent practice in the application of the treaty" shall also be taken into account. Discussing conditions that must be fulfilled before a conventional rule can be considered to have become a rule of customary law, the International Court of Justice declared in the landmark *North Sea Continental Shelf Cases*[21] that, in addition to widespread practice and representative participation in a convention, the practice must include states whose interests are "specially affected." With only the Soviet Union and United States active in outer space before and for some time after the entry into force of the Outer Space Treaty, the "practice" of even one space power, clearly a "specially affected" State, carried substantial weight in law. All the more so when supported by several States with developing capabilities.

At the beginning of negotiations between the US and its potential partners in the international space station, the US DoD was involved in NASA's conceptual work of the space station but lost interest when it could not find a

---

19. As of March 1995, the RAND study identified five categories under which these agreements fall: (1) basic exchange and cooperative agreements between at least ten countries and the Defense Mapping Agency; (2) similar agreements with the DoD (Poland and Hungary); (3) an agreement with New Zealand for installation, operation, and maintenance of Global Sea Level Data Collection Stations; (4) a Memorandum of Understanding with Germany for a maritime control aircraft programme; and, (5) Memoranda of Agreement with Australia and New Zealand specifically concerning the NAVSTAR GPS.

20. (1980) no. 58 United Kingdom Treaty Series, Cmd. 7964. Opened for signature 23 May 1969; entered into force 27 January 1980.

21. (1969) I.C.J. Reports 3, para. 73. A rule becomes a rule of customary international law when a significant majority of states, including states whose interests are especially affected, act in accordance with that rule because they believe it to be binding. In the application of such a rule state practice must be both extensive and virtually uniform.

valid military use. The orbital inclination (28) makes the space station of little value for military purposes.

Originally, DoD wanted to include in the Intergovernmental Agreement (IGA) a provision whereby all partners would have the right to use their allocated shares of space station resources for any purpose, including, in the case of the US, national security.[22] A partner's use of its allocated share would not be subject to objection unless such use would affect the "foreign policy or national security interests of the party providing the element or of the United States."[23] In such an instance, the concurrence of the partner providing the element or the United States would be required. In essence, the US had the right to use all elements of the station for national security purposes. If such use involved another partner's element, that partner could object and prevent the questioned activity. No one however, had the right to object to the US using its own element for national security purposes. The US had the right, however, to veto a partner's use of its own element if the US felt its national security or foreign policy interests were affected.

Needless to say such a proposal did not meet with the partners approval. The IGA adopted and signed in September 1988 retains provisions requiring that the station be operated in accordance with international law.[24] Article 2 further requires that the station be "utilized in accordance with international law, including the Outer Space Treaty, the Rescue and Return Agreement, the Liability Convention and the Registration Convention." Article 9 states that the partner providing an element has the right to determine if the anticipated use of that element complies with the peaceful purposes requirement.[25] In an article dealing with the evolution of the station, language requiring that it remain a civil station operated for peaceful purposes was inserted.[26]

Finally, three separate Exchange of Letters were signed between the partners and the US whereby each partner reiterates that the US has the right to use its elements and the infrastructure for national security purposes so long as such use is for peaceful purposes and in accordance with international law.

---

22. Department of State, An Agreement Between the Government of the United States and the Government of Canada on Cooperation in Detailed Design, Development, Operation and Utilization of the Permanently Manned Space Station (Draft, 3 February 1987).

23. *Idem*, articles 9.4 and 9.5.

24. Agreement among the Government of the United States of America, Governments of Member States of the European Space Agency, the Government of Japan, and the Government of Canada on the Cooperation in the Detailed Design, Development, Operation and Utilization of the Permanently Manned Space Station (hereinafter IGA), entered into force in September 1988, Article 1. For additional information see also, K.K. Spradling, *Military Use of the International Space Station*, unpublished Master of Laws Thesis, McGill University, November 1988.

25. *Idem*, Article 9.8(b).

26. *Idem*, Article 14.1.

Thus, the Exchange of Letters has all three partners acknowledging in effect, the United States' long-standing position that the phrase "peaceful purposes" in international law permits some military activities. While all may have acquiesced to that interpretation in practice by not raising the issue, some have never officially admitted that the American view is correct.

More recently, in January 1998, a new Intergovernmental Agreement was signed between the same parties as the above-mentioned IGA plus the Government of the Russian Federation.[27] The provisions dealing with national security issues and peaceful purposes are identical to those of the original IGA. The parties could have used language banning all military activities on board the civil space station, or banning specific military activities, or list criteria by which certain activities could be judged on a case-by-case basis. The parties to the IGA and 1998 IGA could also have included inspection provisions such as those found in the Antarctic Treaty[28] and the 1979 Moon Agreement.[29] Under the Outer Space Treaty only stations on the Moon or other celestial bodies are subject to on-site inspections.[30]

Principles on Remote Sensing

Discussions on remote sensing within the United Nations started in 1968 at the first United Nations Conference on the Exploration and Use of Outer Space (UNISPACE I). A working group on remote sensing was established and started its work to study a proposal presented to the Committee on Peaceful Uses of Outer Space (COPUOS) by Argentina in 1970. Several draft proposals were discussed in the following sixteen years. The major problems which plagued COPUOS can be summarized as follows:

Permissibility. A group of states primarily from the developing countries and the former Soviet block countries argued that data gathering from outer

---

27. Agreement Among the Government of Canada, Governments of Member States of the European Space Agency, the Government of Japan, the Government of the Russian Federation, and the Government of the United States of America Concerning Cooperation on the Civil International Space Station (hereinafter IGA 1998), signed on 29 January 1998, Washington, D.C. The IGA 1998 supersedes the previous IGA.
28. Article VII(3) provides that "all areas of Antarctica, including all stations, installations and equipment within those areas, and all ships and aircraft at the points of discharging or embarking cargoes of personnel in Antarctica, shall be open at all times to inspection by any observers designated in accordance with paragraph 1 of this article." Antarctic Treaty, (1961) 402 U.N.T.S. 71. Opened for signature on 1 December 1959; entered into force 13 June 1961.
29. Article 15.1 "(a)ll ... stations and installations on the Moon shall be open to other States Parties. Such Sates Parties shall give reasonable advance notice of projected visit, in order that appropriate consultations may be held and that maximum precautions may be taken to assure safety and to avoid interference with normal operations in the facility to be visited."
30. Article XII.

space should be under strict state control and required the prior consent of a given sensed state.

Access. This same group of countries also argued that the sensed state should have priority access to any data pertaining to its territory and natural resources.

Dissemination to third parties. A joint draft presented by Argentina and Brazil required prior authorization of a sensed state before any data was disseminated to third parties. A joint French/Soviet draft proposed prior approval by the sensed state for data having a resolution of better than 50 metres.

In 1986, the "Principles Relating to Remote Sensing of the Earth from Space" were adopted. The eleven principles adopted by consensus constitute the main body of international legal provisions applicable to civil Earth remote sensing. The Principles are laid down in a resolution of the United Nations. A resolution itself is not binding upon states but reflects the opinion of governments at a given point in time. This means that when the principles are concerned with general norms of international law, they provide the basis for the progressive development of law and a more expeditious means of consolidation of customary rules.

In an overview, the most important provisions are:

Principle I

Principle I limits the applicability of the Principles to sensing of the Earth's surface from space for the purpose of improving natural resources management, land use, and the protection of the environment. Thus, no military and meteorological systems are addressed by the Principles. Remote sensing activities are defined as the operation of remote sensing space systems, primary data collection and storage stations, and activities in processing, interpreting, and disseminating the processed data.

Principle II

Principle II reiterates Article I of the Outer Space Treaty but adds "taking into particular consideration the needs of the developing countries." This special attention for developing countries can also be found in the International Telecommunications Union (ITU) provisions with respect to the use of the geostationary orbit.

Principles III and IV

Principles III and IV provide that remote sensing activities shall be carried out in conformity with international law, including the Charter of the United Nations, the provisions of the Outer Space Treaty, and the provisions of the ITU and gives special emphasis to Article I of the Outer Space Treaty *vis-à-vis*, in particular, benefit and interests of all countries and with respect to full and permanent sovereignty of all states and peoples over their own wealth and natural resources. Remote sensing activities may not be conducted in a manner detrimental to the legitimate rights and interests of the sensed state.

Principle XII

Principle XII is the key principle and provides for the nondiscriminatory access by the sensed state on reasonable cost terms, as soon as the primary data and the processed data concerning the territory under its jurisdiction are produced. The sensed state shall also have access to the available analyzed information concerning the territory under its jurisdiction that is in the possession of any state participating in remote sensing activities on the same basis and terms, taking particularly into account the needs and interests of the developing countries.

It allows Earth observation without prior consent of the sensed state but gives the sensed state the right to have access as soon as the data (either primary or processed) are produced within a reasonable time delay. Finally, the Principle not only refers to primary and processed data but also applies to analyzed data in the possession of any state participating in the remote sensing activity. Here also the states under which jurisdiction ground stations are established or where distributors are incorporated are obliged to act in conformity with the provisions of the principles.

Thus, two main principles can be identified from the resolution. The first principle is that there is freedom of collection of data by remote sensing satellites without prior consent of the sensed states. The second is that there is freedom of dissemination of these data, although there is state responsibility when these data cause prejudice to the sensed state.

Remote sensing is allowed under international law and must be carried out in such a way that it will not negatively impact on the interests of other states. With the emerging Very High Resolution Remote Sensing Systems (VHR) following the Clinton policy announced in March 1995 to allow commercial remote sensing systems with a resolution of up to one metre, a number of legal issues came to the fore, notably with respect to issues of national security. Until 1995, there was a consensus within the state community not to commercially distribute data with a resolution higher than 10 metres. This is, for example, the official French policy with respect to SPOT satellites. The commercial VHR systems, all private US ventures, will *de facto* end this consensus.

The commercial VHR remote sensing systems illustrate a general weakness of international law — namely, that licensing of space activities is purely a national prerogative within the limits of international law and space law principles, but the impact of these (national) license policies on the international level can have very serious repercussions. The satellite S-PCS systems example illustrates the concern of other states that the US with its policies will prevent other states from engaging in similar activities at a later time. There is also the concern of states that they will be unable to control services within their own territory. For example, when VHR systems are operational and working, a nationally-imposed limit on the resolution of data concerning its territory does not make sense when this same data can be acquired from a neighboring state. Some states, therefore, are asking for the better coordination of the various national policies, whereas other states, notably the US, will attempt to create as few obstacles as possible for this new economic activity.

Missile Technology Control Regime

Efforts by the international community to control ballistic missile proliferation focus on the Missile Technology Control Regime. Initially drafted by a group of seven states in 1987,[31] the MTCR currently has close to 30 members. The fundamental assumptions upon which the MTCR guidelines were drafted are:

- there is a "core" of controllable technology;
- without access to this "core" technology, third parties cannot develop missile technology indigenously; and,
- the cost of circumventing the control regime is enough to deter "problem" countries from making the effort.

The MTCR is not a treaty but a set of guidelines to limit the conditions under which missile technology may be transferred. The guidelines do not prohibit exports but require governments to judge whether specified items meet a series of criteria before being approved for export. The MTCR places strict controls on exports of space-launch vehicles, components, and products of technologies used in civilian space programmes. Therefore, there is a shift from controlling nuclear weapons to delivery systems. The MTCR also includes an Annex listing technologies to be controlled and an informal mechanism by which the partner states can share information about potential transfers.[32]

Though the regime has successfully slowed the overall rate of ballistic missile technology proliferation, it still faces many challenges, such as: the failure of major suppliers to join the regime; the growing sophistication of production capability in potential suppliers, who also have not joined the MTCR; the increased risk of proliferation stemming from the weakness of enforceable export controls in the states of the former Soviet Union; and,

---

31. The original participants of MTCR were Canada, France, Germany, Italy, Japan, the United Kingdom, and the United States.

32. Category I includes: complete rocket systems (including ballistic missiles and space-launch vehicles) of a certain range. This category now also includes chemical, biological, and nuclear systems of any range. Category II includes a long list of dual-use technology such as communication satellites and propulsion components.

above all, the fundamental inability of any supply-side control to halt proliferation.

If the MTCR is to be a more effective non-proliferation regime, it must evolve from an export control regime to a broader multilateral non-proliferation arrangement that develops and promotes international norms in transfer and control of missile technology. In this context, France has stated that the MTCR:

> ... should only be a stage towards a more general agreement, one that is geographically more extensive, better controlled and applicable to all. The agreement would lay down rules promoting civilian co-operation in space, while removing the dangers of the diversion of technology for a military ballistic capability. Here again, the aim would be to arrive at a situation where all States wishing to gain access to space for development purposes would cooperate in a framework guaranteeing security.[33]

The negotiation of a more formal agreement would, however, place greater emphasis on prohibition. Explicit prohibitions would be difficult to agree on amongst the MTCR members, many of whom have considerable commercial interest in exporting systems and technologies for civilian space programmes.

Furthermore, consideration will have to be given on how to convince key suppliers and potential suppliers to abide by the non-proliferation objectives underlying the regime, including the best way to isolate states that persist in getting a missile delivery capability for weapons of mass destruction. Based on the 1968 Non-Proliferation Treaty[34] experience, some of the more worrisome countries will undoubtedly resist MTCR limitations by pointing to the inequality of an arrangement that preserves missiles for a few states while denying them to others.

In March 1992, the US imposed sanctions against the Russian entity GLAVKOSMOS and the Indian Space Research Organization (ISRO). The US, along with other MTCR members, considered that the proposed sale violated MTCR guidelines. As stated in the US Freedom Support Act (S2532), aid cannot be provided to Russia or any other republic if it violates the MTCR.[35] Such linkages between economic and security issues can help control access to space technology.

---

33. "Letter dated 3 June 1991 from the Representative of France Addressed to the President of the Conference on Disarmament Transmitting the Text of Arms Control and Disarmament Plan Submitted by France on 3 June 1991," CD/1079 (1991).
34. Treaty on the Non-Proliferation of Nuclear Weapons, (1970) 729 U.N.T.S. 161. Opened for signature 11 February 1971; entered into force on 18 May 1972.
35. See M.S. Smith, "Buying Russian Technology — Pros and Cons for the US Programme," Reports, *Space Policy*, November 1992, 362.

Finally, another means of improving on the MTCR is to strengthen restrictions in terms of national legislation, which could lead to a better coordination among member states and improved controls. It should be recalled that the guidelines depend on national legislation for their legal validity. A number of countries, notably the US and Germany, are attempting to strengthen national control legislation.[36]

Another proposal which deals specifically with the regulation of dual-use technology was presented in 1991 in a joint Argentinian/Brazilian paper at the United Nations Disarmament Commission (UNDC).[37] This paper urges the international community to establish a set of basic rules for the international transfer of sensitive technologies. The Draft Guidelines on the International Transfer of Dual-Use Technology aims at ensuring the flow of dual-purpose technology for peaceful purposes via the establishment of a mechanism which would be transparent, universal, and effectively controlled through an international system of safeguards. The Draft contains no verification mechanisms.

More recently, the Legal Subcommittee (LSC) of COPUOS has been discussing a working paper tabled by a group of developing countries dealing with "Principles Regarding International Cooperation in the Exploration and Utilization of Outer Space for Peaceful Purposes."[38] This Draft Set of Principles was rich in New International Economic Order language aiming at what could only be described as "forced" cooperation and an automated transfer of financial and technological resources from North to South. This came as a surprise to the industrialized countries who had not expected such a strong attack on the freedom of space exploration and utilization.

Making it clear that numerous developing countries already participated in bilateral space projects and that industrialized countries do provide substantial amounts of funding for building satellite utilization capacities in developing countries (through programmes such as the UN Space Applications Programme), the industrialized countries set a new tone for the debate: industrialized countries are already making efforts voluntarily and cooperation must be beneficial to both sides.

---

36. In its 1994 National Defence Authorization Act, the US confirms a strict interpretation to treat the sale of launch vehicles as restricting any sale of missile technology. The US policy is not to increase the number of nations acquiring space-launch vehicles. For further discussion on this issue see, P. Gasparini Alves, *Outer Space Technologies and International Security* (Geneva: UNIDIR, 1991), 116–19.

37. See, "The Role of Science and Technology in the Context of International Security, Disarmament and Other Related Fields," International Transfer of Sensitive Technology, Working Paper submitted by Argentina and Brazil, Disarmament Commission, A/CN, 10/145 (1991).

38. Working paper presented by Argentina, Brazil, Chile, Columbia, Mexico, Nigeria, Pakistan, The Philippines, Uruguay, and Venezuela. UN Doc. A/AC.105/C.2/L.182 of 9 April 1991.

Thus, Germany and France decided to present a working paper to the 1995 LSC — a working paper containing a draft "Declaration on International Cooperation in the Exploration and Use of Outer Space for the Benefit and in the Interests of all States Taking into Particular Account the Needs of Developing Countries."[39] This working paper rested on two considerations. First, that states are free to determine all aspects of their international cooperation, whether bilateral or multilateral, or whether commercial or noncommercial, including development cooperation. Second, that states should choose the most effective and appropriate mode of cooperation in order to allocate resources efficiently.

Remarkably, a merger between both sets of draft Principles occurred in 1996. The final text of the Declaration was adopted by the UN General Assembly as Resolution 51/122 of 13 December 1996. As with all UN adopted Declarations, this set of Principles is non-binding and simply reflects the thinking of governments on this issue at a given point in time.

This Declaration confirms the freedom of exploration and utilization of outer space while reminding the space powers — in a productive manner — to fulfil their obligation to conduct space activities for the benefit of all countries, which means that the space powers should foster international cooperation on an equitable and mutually acceptable basis. This, in turn, means that developing countries now have higher incentives to put their energies into a "demand-pull," which should be well prepared by their own efforts. This will make many of them more equal partners, which the space powers are willing to accept.

One could imagine a similar approach being adopted within the CD, notably with respect to remote sensing. Numerous CD member states either have their own remote sensing satellites or benefit from the use of remote sensing data for both civilian and military purposes. A review of the huge proliferation of remote sensing satellites and the use by CD member states of such data would be a valuable point to make. This could then be coupled with the fact that the UN Principles on Remote Sensing, which were adopted by consensus, have been followed by all states. No formal protests or objections to these principles have ever been made.

Outer Space Treaty

The Treaty on Principles Governing the Activities of States in the Exploration and Use of Outer Space Including the Moon and Other Celestial Bodies,[40] commonly referred to as the Outer Space Treaty, is considered the cornerstone international space law treaty.

---

39. UN Doc. A/AC.105/C.2/L.197 of 27 March 1995.

40. See *Supra*, note 1.

Article IV contains the only provision of the Outer Space Treaty addressed specifically to military activities and reads as follows:

> States Parties to the Treaty undertake not to place in orbit around the Earth any objects carrying nuclear weapons or any other kinds of weapons of mass destruction, install such weapons on celestial bodies, or station such weapons in outer space in any other manner.
>
> The moon and other celestial bodies shall be used by all States Parties to the Treaty exclusively for peaceful purposes. The establishment of military bases, installations and fortifications, the testing of any type of weapons and the conduct of military manoeuvres on celestial bodies shall be forbidden. The use of military personnel for scientific research or for any other peaceful purposes shall not be prohibited. The use of any equipment or facility necessary for peaceful exploration of the moon and other celestial bodies shall also not be prohibited.

Paragraph 1, while applicable to all of the space environment, prohibits the presence in these areas of only nuclear weapons and weapons of mass destruction. Furthermore, as is clear from the text of paragraph 2, the area of outer space *per se* is excluded from the requirement of use exclusively for peaceful purposes. The conclusion is inescapable that Article IV of the Outer Space Treaty was, at best, a first step toward controlling military uses of outer space.

A memorandum suggesting a widening of the scope of the Outer Space Treaty's provisions was submitted by Italy.[41] This proposal was largely inspired by the developments in space technology at that time, notably development of weapons such as interceptor/destructor and anti-satellite weapons. The Italian document suggested a total ban on such military activities as the development and use of Earth- or space-based systems designed to damage, destroy, or interfere with operations of other States' satellites.[42] It also put forward a draft Protocol called Additional Protocol, to the Treaty. Article I of this draft reads as follows:

> Outer space, including the moon and other celestial bodies, shall be used for peaceful purposes only. States Parties to this protocol undertake to refrain from engaging in, encouraging or authorizing, directly or indirectly, or in any way participating in any measures of military or other hostile nature, such as the establishment of military bases, installations and fortifications, the stationing of devices having the same effect, the launching into Earth orbit or

---

41. "Additional Protocol to the 1967 Treaty on Principles Governing the Activities of States in the Exploration and Use of Outer Space, Including the Moon and Other Celestial Bodies with a view to Preventing an Arms Race in Outer Space," CD/9, 26 March 1979.

42. *Ibid*, 2.

> beyond of objects carrying weapons of mass destruction or any other types of devices designed for offensive purposes, the conduct of military manoeuvres, as well as the testing of any type of weapons.[43]

Venezuela and Peru also made proposals at the CD to modify the Outer Space Treaty.[44] The Venezuelan delegate argued that a simple amendment would suffice to turn the Treaty's partial prohibition into a total ban by adding the words "or any type of space weapons" to Article IV.[45] Venezuela also proposed the insertion of a new paragraph under which states parties would undertake not to develop, produce, store, or use space weapons.

These initiatives raised the question as to where their discussion and negotiation should take place — at the CD itself or at COPUOS, the latter being the forum where the Treaty was originally negotiated. This remains a controversial issue. Some delegates favoured the choice of the CD because the subject concerns international security. Others felt strongly that any changes to an instrument negotiated at COPUOS should be carried out in the same negotiating forum.

The year 1997 marked the 30th anniversary of the entry-into-force of the Outer Space Treaty. The issue of amending the Treaty was raised. There was an agreement to start a multi-year review of the status of ratification of the five legal instruments of COPUOS. However, it was also explicitly stated that this will not lead to any amendment of the treaties.[46]

It should also be noted that, in fact, for the non-weaponization of outer space to become a reality, a new international agreement is not needed. All that would need to be done is to amend Article IV (1) of the Outer Space Treaty to read:

> States Parties to the Treaty undertake not to place *in outer space* [instead of "in orbit around the earth"] any objects carrying nuclear weapons or any other kinds of weapons of mass destruction, install such weapons on *the moon or* on celestial bodies, or station *weapons* [instead of "such weapons"] in outer space in any other manner.[47]

The removal of the word "such" would mean that *no* weapon could be placed in earth orbit, while the other changes would ensure the non-

---

43. *Ibid*, Annex I, 1.

44. For Peru see "Proposal for Amendment of the Treaty on Principles Governing the Activities of States in the Exploration and Use of Outer Space, including the Moon and Other Celestial Bodies," CD/939, 28 July 1989. For Venezuela, see CD/851, 2 August 1988.

45. CD/PV 398, 9.

46. Information obtained in a conversation with someone from the United Nations Office of Outer Space Affairs, Vienna, February 1997.

47. See Christol, *supra*, note 10, 846–47.

weaponization of outer space. In view of the above-mentioned comment made before COPUOS in 1997, it does not appear that such an amendment is likely to be discussed, let alone made.

Recommendations

1. Within the CD, and even COPUOS, urge states to invoke and make effective use of existing treaty provisions such as:
   (a) Article IX of the Outer Space Treaty provides that a state party to the Treaty "which has reason to believe that an activity or experiment planned by another State Party in outer space, including the moon and other celestial bodies, would cause potentially harmful interference with activities in the peaceful exploration and use of outer space, including the moon and other celestial bodies, may request consultation concerning the activity or experiment." To date, this Article has never been invoked or used.
   (b) Article IV of the Registration Convention urges states to provide more timely and precise information about the launch of space objects and to update the information. Ideally, states should provide information concerning launches *prior* to the actual launch, rather than "in a timely fashion" as currently stated in the above-mentioned article.
   (c) Article VI of the Outer Space Treaty, prescribing international responsibility of States for activities of non-governmental entities having their nationality, merits another look with a view toward making the supervisory duties of a state more precise to ensure the orderly and safe use of the space environment. This is especially important in view of future commercial trends in the areas of both telecommunications (mobile satellites with their constellation of satellites) and Very High Resolution Systems of remote sensing.
2. The ABM Treaty, SALT I and II, the START Treaties, and other more recent arms control agreements have legitimized the use of national technical means (NTM) of verification, as well as the principle of non-interference with such satellites. The CD could investigate the possibility of widening the scope of protection of certain satellites beyond the bilateral level.
3. Present before the CD an overall review of current military activities carried out in outer space, highlighting in particular the stabilizing role played by military satellites (communications, navigation, and reconnaissance, amongst other functions). The review should also include the proliferation of commercial uses of outer space which have potential military value, notably in the area of remote sensing. State practice in terms of use and dissemination of remote sensing data should then be made, along with a reminder of the current body of international law dealing with remote sensing.

4. Explore the possibility of having some sort of informal working group set up between members of the COPUOS Scientific and Technical Subcommittee and the CD. Members of this joint working group could attend deliberations in both bodies as observers.
5. Ensure that both background papers and final proceedings of the 1999 UNISPACE III conference, organized by COPUOS and to be held in Vienna, be fed into the CD.
6. Create a CD-ROM and also put out on the Internet the main issues concerning the prevention of an arms race in outer space. This is important to inform people of the current situation and future scenarios, raise public awareness, and stimulate debate and discussion worldwide.
7. Continue to advance and reiterate suggestions made in past Canadian working papers to the CD with respect to confidence building measures.

Chapter Six

A Variant of a Mandate for an Ad Hoc Committee on Outer Space within the Conference on Disarmament: A Convention for the Non-Weaponization of Outer Space

Phillip J. Baines*

Introduction

An earlier Conference on Disarmament (CD) mandate for outer space — namely, the Prevention of an Arms Race in Outer Space — requires revision to reflect realities of the new international security concerns of the post-Cold War world. Choosing to ignore the presence of the word "prevention" in the current mandate, a lone voice in the CD has claimed that there never has been, nor is there currently, an arms race in outer space. But as the CD operates by consensus, the international community has yielded to this isolated position in an attempt to flush out the last remaining holdout for progress in security issues pertaining to outer space. The last UN General Assembly therefore passed, unopposed, A/RES/52/37 calling for the CD "to re-examine the mandate ... for the re-establishment of the Ad Hoc Committee ..." for outer space.[1]

Recognizing the opportunity to promote a long-standing Canadian objective, Canada has proposed that an ad hoc committee be formed with the mandate to negotiate a convention for the non-weaponization of outer space.[2] The advance of military technology, policy, and doctrine for outer space poses a clear and present danger to international peace and security in the first decade of the next millennium. This is made all the more acute by a certain

---

* This chapter is an edited version of a paper commissioned for the Fifteenth Annual Ottawa NACD Verification Symposium, "Arms Control and the Rule of Law: A Framework for Peace and Security in Outer Space," 11–14 March 1998, Montebello, Quebec, Canada. The opinions expressed in this chapter are those of the author and do not necessarily reflect those of the Department of Foreign Affairs and International Trade or the Government of Canada.

1. United Nations Resolution A/RES/52/37, "Prevention of an Arms Race in Outer Space," New York, 23 December 1997.
2. Conference on Disarmament Document CD/1487, "Canada Working Paper Concerning CD Action on Outer Space," Geneva, 21 January 1998.

government in waiting urging abrogation of the ABM Treaty. Should weapons ever be placed into outer space, all objects in outer space, or on or above the surface of the Earth, could be threatened by destruction at the speed of light. Space-based lasers or conventional missile interceptors capable of ballistic missile defence would guarantee the presence of nuclear weapons on Earth in perpetuity as sufficient numbers of weapons necessary to saturate these defences would be required to maintain credible nuclear deterrents. The lack of progress which this would entail on the nuclear disarmament side of the Non-Proliferation Treaty could then result in a complete rupture of the non-proliferation side of that Treaty by nations that hold the current regime as discriminatory. This "weaponization" of outer space will also disrupt further international cooperation in outer space at the critical juncture when humanity's attention turns from the exploration of outer space to the exploitation of outer space. These economic interests may further exacerbate the international peace and security outlook for outer space early in the next millennium. This chapter presents the case for a convention for the non-weaponization of outer space. It concludes that such a convention is equitable, verifiable, and in the national security interests of all nations.

The Current Legal Regime for Outer Space

The current legal regime for outer space grew out of a patchwork of international conventions and resolutions implemented since the first artificial satellite orbited the Earth in 1957. Today, no single treaty fully specifies which activities in outer space are permitted and which are prohibited. The cornerstone of space law, the Outer Space Treaty, requires, *inter alia*, that outer space simply be used for "peaceful purposes." The Outer Space Treaty also requires that celestial bodies including the Moon shall be used "exclusively for peaceful purposes." Both of these dated expressions have grown to become codifications of ambiguity rather than eruditions of clarity with time. For example, some jurists claim that "peaceful uses" of outer space include uses for "national security." In this sense, national security becomes a further obfuscation for the military use of space-based weapons. Others maintain that "peaceful purposes" excludes any "military uses" whatsoever. In an earlier Canadian paper, usage within the CD holds that "peaceful purposes" includes "military uses" of a "nonaggressive" nature, whereas "exclusively peaceful purposes" excludes all "military uses," whether "aggressive" or "nonaggressive."[3] Lying somewhere in between these two extreme positions is the concept of "weaponization" that distinguishes "military force support" from "military force application" roles of current military doctrine.

The legal regime for outer space is further fractured by the fact that not all existing agreements relating to outer space are open to accession by all countries of the United Nations system. Substantial contributions to the legal

---

3. Conference on Disarmament Document CD/716 or CD/OS/WP.15, "Terminology Relevant to Arms Control and Outer Space," Canada Working Paper, 16 July 1986.

regime for outer space have been negotiated exclusively by the United States and the former Soviet Union. Notably, the ABM Treaty applies to a mere five states, whereas the Outer Space Treaty has 120 signatories. Consequently, opportunities exist for nations to develop, test, and deploy ballistic missile defences in outer space, and moreover, key nations possess national technical means of verification that are not granted the same level of protection against interference as either American or Russian systems. The international community can no longer expect that further collusion by these two nations to act in their limited security interests will not occur at the expense of wider international security concerns, nor, expect that domestic forces within these nations can maintain their commitment to the successful management of their strategic relationship in a multipolar world, except through participation in additional multilateral agreements as negotiated by the Conference on Disarmament.

The legal delineation between airspace, which is subject to claims of national sovereignty, and outer space, which is not subject to claims of national sovereignty, remains unresolved since the 4 October 1957 launch of *Sputnik*.[4] Technological progress since the 1950s has increasingly blurred the distinction between airplanes and rockets. The first flight of the X-15 rocket plane created a new aerospace object that could operate at the highest reaches of the atmosphere and the lowest reaches of outer space. Both the US space shuttle and the Russian *Buran* space shuttle shift from rocket to satellite to aircraft at some point in their respective flight regimes. Next generation single-stage-to-orbit reusable launch vehicles with lifting body shapes will possess the capability to cruise in the upper reaches of the atmosphere more like an airplane than like an orbiting satellite. As a consequence, aerospace objects will pose additional challenges to air and space law. The absence of legal measures may well spillover to international security concerns if nations introduce ballistic missile defences capable of serving anti-satellite roles and as nations fear about whether these new aerospace objects carry commercial satellites or weapons of mass destruction on fractional orbit, ballistic, or hypersonic flight trajectories.

The "province of all mankind" provisions of the Outer Space Treaty and the Moon Treaty have not been universally accepted as the discrepancy in the number of signatories of each treaty intimates. The inability of the international community to agree on mechanisms for the management of extraterrestrial resources will become increasingly problematic in the next millennium. As humanity launches the international space station to establish a permanent presence in outer space, attention will increasingly turn from exploring outer space to exploiting outer space. In the absence of a regulatory environment to adjudicate disputes, competitive pressures may lead to international conflict.

---

4. Report of the Legal Subcommittee on the Work of Its Thirty-Sixth Session (1–8 April 1997), United Nations Committee on the Peaceful Uses of Outer Space, Doc. A/AC.105/674, Vienna, 14 April 1997.

Should any nation first station weapons in outer space, the freedom of use provisions of the Outer Space Treaty would vanish with the implicit approval of that nation. Furthermore, weapons in outer space could deny any nation's access to outer space by threatening space-launch vehicles and presenting a gauntlet to spacecraft returning with extraterrestrial resources. Space-based weapons could, therefore, effectively permit a single nation to usurp the wealth of the universe.

The core body of work that defines the arms control and disarmament dimensions of outer space include: the Charter of the United Nations, the Partial or Limited and Comprehensive Test Ban Treaties, the Outer Space Treaty, the Strategic Arms Limitation Talks I and II Agreements, the Anti-Ballistic Missile (ABM) Treaty, and Protocol to the ABM Treaty. A host of other non-proliferation, arms control, and disarmament treaties, agreements, and resolutions make significant contributions to the current legal regime of outer space through direct reference to that domain (i.e., Environmental Modification Convention), or through establishing legal precedents or international norms that may contribute to fashioning law applicable to outer space (i.e., the Antarctic Treaty[5]).

Using the tenet of international law that anything not specifically prohibited is permitted, an analysis of these core documents would conclude that the international community has successfully prohibited the testing, deployment, and use of space-based weapons of mass destruction. However, that same analysis would conclude that the multilateral outer space legal regime has thus far failed to prohibit the testing, deployment and use of the equally threatening space-based conventional weapons using chemical explosives and kinetic energy impacts. Moreover, the same conclusion can be drawn for space-based "exotic" weapons based on "other physical principles" like lasers, neutral particle beams, and high-power microwave beams. Further analysis of the outer space legal regime will also conclude that a legal vacuum with regard to Earth-based nuclear, conventional, and "exotic" weapons that directly ascend, "pop-up," loiter, or otherwise "reach into" the domain of outer space are also not prohibited. Additionally, all Earth-based weapons are still free to transit through outer space on their ballistic missile or fractional orbit bombardment system trajectories. The penchant for technological progress to outstrip informed policy has become so acute that a former prime minister of Canada called the Outer Space Treaty "patently inadequate" a mere fifteen years after entry into force.[6]

---

5. The Antarctic Treaty (1959) is often referred to in outer space discussions as an example of demilitarization. The Treaty uses the expression "exclusively for peaceful purposes" as a coda prohibiting military use.

6. "Technological Momentum the Fuel that Feeds the Nuclear Arms Race," An Address to the Second United Nations Special Session on Disarmament by the Right Honourable P. E. Trudeau, Prime Minister of Canada, New York, 18 June 1982.

Why is Outer Space a Multilateral Concern?

The rationales for multilateral participation in the definition of an adequate arms control regime for outer space are myriad. The foremost is the spatial relationship between outer space and the countries of Earth. Another is the increasing plurality of activities in outer space. Yet more importantly, is the matter of obligation by the international community.

All nations border on outer space. The proximity of outer space to each and the speed-of-light-approach with which space-based weapons could strike forces each nation's national security assessment to consider the intent, military capability, and actions of its neighbours everywhere on the globe. A de-orbiting weapon would take but a few minutes to fall from the lowest reaches of outer space, while a space-based directed energy weapon requires less than one millisecond to strike an air-based object flying at an altitude of 10 kilometres. Ambassador Carasales of Argentina summed it up most eloquently when he said, "It is precisely to gain mastery over the Earth that attempts are made to control space, as in the past the aim was to control the seas in order to have dominion over the continents."[7] The community of nations also can ill-afford the risk of a single nation exercising control over the wealth of the universe. Unrestrained weaponization of outer space has the potential to turn low Earth orbits into the "Dardanelles of the cosmic seas" and the "ultimate high ground" well within the first decade of the next millennium.

The plurality of outer space activity has grown significantly since 1967, when the Outer Space Treaty was ratified. Whereas in 1967 outer space was essentially the exclusive domain of a few states, today practically every state benefits from it, over 30 countries possess significant space industries, and eight nations provide access to outer space through space-launch vehicles. Whereas the "space race" was a competition between two nations, today's environment is characterized by a competition in a marketplace of enterprises. For example, space-launch vehicles are now offered on a commercial basis, as opposed to through the government-to-government agreements of an earlier era. Joint ventures further transform this market of competitors into an oligopoly of collaborators. A prime example is the Sea Launch Company which combines aerospace giants in Russia and the United States to provide access to space from floating equatorial platforms.[8] Investment in satellite operations has also become more international where equity capital from all quarters of the globe provides the investment for the Iridium mobile satellite system.[9] Through rapid commercialization, outer space is beginning to live up to the

---

7. Ambassador Carasales of Argentina, Conference on Disarmament Doc. CD/PV.253, Geneva, 27 March 1984, 20–23.
8. P. Sparaco, "Space Launch Industry Faces Dramatic Changes," *Aviation Week and Space Technology*, 145:25 (16 December 1996), 86.
9. International consortium composition of Iridium is available at http://www.iridium.com.

res communal environment first fashioned by our forebears in international law.

Persons representing several additional nations have since joined the early pioneers of Soviet Cosmonauts and American Astronauts in the exploration of outer space. The imminent launch of the international space station foretells of increased international cooperation in outer space. It is hoped that the station may be followed by other variants *ad infinitum*. A marked transition from the exploration of outer space to the exploitation of outer space will occur as humanity establishes a permanent presence there. And yet, the international community has still to provide a secure and stable environment in outer space free of weapons which could destroy such investments in the blink of an eye. A known and stable security and legal regime for outer space is a prerequisite for increased commercial investments in enterprises that may include ballistic transoceanic passenger aerospace flight or other exciting prospects like extraterrestrial mining on the Moon or other celestial bodies.

Finally, outer space is a multilateral concern through the construct of the United Nations. The purpose of the United Nations, as defined in Article 1 of its Charter, is "To maintain international peace and security, and to that end: to take effective collective measures for the prevention and removal of threats to the peace ..." Arms control agreements negotiated in the CD, defined as the "single multilateral negotiating forum,"[10] constitute but one effective collective measure to remove the threat that the weaponization of outer space would entail for the international community's peaceful use of outer space. Article VI of the Non-Proliferation Treaty further obliges State Parties to negotiate "A treaty on general and complete disarmament under strict and effective international control." In this vein, UN/RES/39/59, which passed unopposed with 140 yeas, zero nays, and one abstention, expressed the international will that Article VI of the NPT applies equally to the domain of outer space.[11] In the year subsequent to this definitive resolution, the Conference on Disarmament first commissioned an Ad Hoc Committee for the Prevention of an Arms Race in Outer Space. Work begun in the clenches of the Cold War to prevent the migration of conflict into outer space remains unfinished in the post-Cold War's transition to a "*nouveau age d'or*." To paraphrase a former US President, "we must accomplish in peace what eluded us in war." The international community, and the Conference on Disarmament in particular, has the obligation to secure the domain of outer space from the threats to humanity posed by space-based weapons.

---

10. Paragraph 120 of Resolution S-10/2 of the Final Document of the Tenth Special Session of the General Assembly of the United Nations Devoted to Disarmament, (1978) also known as UNSSOD I.

11. A/RES/39/59, "Prevention of an Arms Race in Outer Space," New York, 12 December 1984, states in part: "Reaffirms that general and complete disarmament under effective international control warrants that outer space shall be used exclusively for peaceful purposes and that it shall not become an area for an arms race."

Current Military Use and Plans for Outer Space

The "stabilizing" military use of outer space, like the "nonaggressive" missions of communication, meteorology, navigation, reconnaissance, and early warning, continue their vigilance today and contribute to the enhancement of international peace and security. Within these military force support missions, the evidence of the world's transformation from an old political world order to a new economic world order can be seen. It is both a matter of industrial Darwinism to maintain aerospace military capacity in the face of Cold War reductions, and an attempt by a single nation to control access to outer space by market domination, that prompts this transition. As a matter of United States policy, these dedicated military missions of the Cold War are increasingly being transferred to civilian ownership. Likewise, non-space-capable nations are exploiting the military utility of access to increasingly sophisticated civilian technology. This aspect is contained in the term "dual use." Dual use expresses the blurring of distinction between military and civilian use and enemy from friendly satellites. A number of examples will serve to illustrate this trend.

The US Air Force's Defense Meteorological Satellite Program (DMSP) is being merged with the US Commerce Department's National Oceanic and Atmospheric Administration's (NOAA) current system to produce a single National Polar-Orbiting Operational Environmental Satellite System (NPOESS) serving all former users.[12] Several nations will have unimpeded access to this essential service for civilian and military use. In the communications domain, a current military commander forecast that 80 per cent of the US military's communications traffic would be better served by civilian satellites[13] is matched with announcements that US military units are buying commercial mobile satellite handsets.[14] The US National Imagery and Mapping Agency (NIMA) will probably also use the services of new 1m resolution Earth remote sensing satellite systems offered by the American Space Imaging Company, EarthWatch Incorporated, and OrbView Incorporated satellites to offset tasking of limited resources with better capabilities.[15] In addition, no one could have foretold the growing dependency civilians would have on the military Global Positioning System (GPS) for navigation, nor the expectation that augmented use of the GPS system would extend to civilian air traffic navigation and air traffic control. During the Gulf War both Iraqi and Coalition forces were using the same civilian ARABSAT satellite for military

---

12. "$149 Million in Contracts Awarded for Environmental Satellite Instruments," NOAA Press Release 97-45, 31 July 1997.
13. Keynote Address by General Thomas S. Moorman, Jr., USAF Vice Chief of Staff at the First National Space Forum held 4–5 June 1997, Washington, D.C.
14. W. Ferster, "U.S. Navy Plans to Buy Additional UHF System," *Defense News*, 12:43 (27 October–2 November 1997), 4, 18.
15. L. David, "Commercial Remote Sensing May Aid Intelligence Imagery," *Defense News*, 12:24 (16–22 June 1997), 62.

communications. This example, better than any other, illustrates the blurring of friendly and enemy satellites and military and civilian use.

It is therefore not a question of banning these civilian or military systems that may support military force application from outer space, but the necessity to ban the "unstable" systems that will exercise the application of military force in or from outer space. Thus, the more "aggressive" uses of outer space that encompass the application of force directly in or from outer space become the paramount concern to ensure international peace and security in the next millennium. Weaponry delineated along the function lines of anti-satellites (ASAT), ballistic missile defence (BMD), or "space-strike" weapons (SSW),[16] are the destabilizing weapons that require prohibition by the international community.

Historically, the ASAT and BMD weapon systems have been researched, developed, tested, or deployed in one form or another by the two Cold War superpowers. Before the relevant restrictions posed by early arms control agreements, it can be recalled that both the United States and Russia had tested and deployed nuclear-tipped ASAT and ABM systems. All of the manifested weapons systems have thus far been Earth-based. The Strategic Defense Initiative (SDI) and Global Protection Against Limited Strikes (GPALS) systems, however, researched several components for weaponry based on conventional and "other physical principles" for space-based deployments. Soviet developments were often cited to justify such efforts in the West. Research, however, is a leading indicator for development, just as development is a leading indicator for testing, testing for deployment, and deployment for use. Today both the United States and the Russian Federation continue research into high-powered lasers for space-based applications. The former Soviet Union's own test programme for advanced high-powered laser research spurred the American response with development of the ground-based Mid-Infrared Advanced Chemical Laser (MIRACL) and the precursors to the current space-based laser programme Alpha administered by the US Ballistic Missile Defense Organization (BMDO).[17] A flight test of a "demonstrator" for the Alpha system could take place as early as 2005.[18] During the 1980s, the US Department of Defense's publications projected that the Soviets may have been ready to deploy similar space-based lasers in the 1990s.[19]

With the test firing of the land-based MIRACL laser against a retiring scientific satellite, ostensibly for determining the vulnerability of such satel-

---

16. The term "space-strike" is an extension of the term "air-strike" to denote space-based weapons attacking Earth-based targets.
17. Limited technical and programmatic information for all US BMD system activities may be found at http://www.acq.osd.mil/bmdo/bmdolink/html/programs.html.
18. J. Erlich, "BMDO, Air Force Boost Laser," *Defense News*, 12:40 (6–12 October 1997), 4.
19. United States, Department of Defense, *Soviet Military Power*, 6th edition (Washington: US Government Printing Office, March 1987), 52.

lites against similar attacks, the United States resumed ASAT testing that had remained dormant since 1985. The last American ASAT test occurred with the air-based Miniature Vehicle (MV). Despite the successful test of the Miniature Vehicle, this system was never operationally deployed. US President Clinton has since exercised his line item veto right to delete funding for the United States' dedicated KE ASAT system that was to have seen 10 systems deployed by the year 2000.[20] The US Congress, however, vows to return funding for this controversial programme. The Russians continue to respect an earlier Soviet test moratorium on ASATs and have openly called for bilateral talks on high-altitude anti-satellite systems.[21] Many argue that, by virtue of disuse, the Soviet era co-orbital system should no longer continue to be considered an operational ASAT capability.

The perceived threat posed by the continued proliferation of ballistic missiles and weapons of mass destruction by nations hostile to the United States and Russia propels them to develop theatre missile defence (TMD) systems. Russia continues to improve its S-300V system to counteract theatre-range ballistic missiles. The United States continues the development of five systems to combat similar threats. US Army systems, known as Patriot PAC-3 and THAAD, are mobile land-based systems using endo- and exo-atmospheric interceptor missiles. The US Navy Theater-Wide and Navy Area-Wide systems are sea-based systems using endo- and exo-atmospheric interceptor missiles. The Air Force's Airborne Laser Program is an air-based system using a directed energy weapon to attack theatre-range ballistic missiles during the boost phase of their trajectories. All of these systems are under the research and development authority of the US Ballistic Missile Defense Organization. It is pertinent that the higher velocity, exo-atmospheric TMD missile systems and the Airborne Laser Program systems may have an inherent ASAT capability against low orbiting satellites.

The Russian Federation continues to operate its single anti-ballistic missile site permitted by the ABM Treaty around Moscow. The United States, after abandoning its earlier Safeguard system at Grand Forks, is continuing its research and development of an ABM Treaty-compliant National Missile Defense (NMD) system. In what is referred to as a "3+3" programme, research and development will be performed in the first three years to enable a decision to deploy which would see operations within three years. Such a decision to deploy would require both a demonstrated technological solution to a difficult problem and an identifiable threat forecast for North America within the three-year implementation horizon. The earliest date for an NMD deployment decision is the year 2000, for a putative deployment in 2003. As the capability of any NMD system must combat ICBM reentry vehicles, an

---

20. W. Ferster, "Budget Battle Brews Over US Army KE ASAT," *Defense News*, 12:11 (17–23 March 1997), 3.

21. B. Gertz, "Yeltsin Letter Reveals Anti-Satellite Weapons," *Washington Times*, 7 November 1997, A9.

NMD interceptor missile will likely have a significant ability to attack satellites in low Earth orbits.

Development of new civilian technologies, such as single-stage-to-orbit reusable launch vehicles and civilian reentry capsules for space science micro-gravity research, may raise security concerns about the possible development of "space-strike" weapons. The ability of the single-stage-to-orbit reusable launch vehicle to glide for extended periods in the upper reaches of the atmosphere, and presumably to fly under power, makes the prospect of military spaceplanes that much more tangible. The inability of the international community to inspect the payload bays of these vehicles prior to liftoff could generate concern that these vehicles may carry weapons of mass destruction, bunker-busting conventional weapons, or internal lasers for BMD or ASAT purposes. The US Air Force, in advocating for military spaceplanes to perform reconnaissance, bombing, and future air lift missions, must be seen as more than a bid to rest funding from US Navy carrier programmes.[22] Another interesting civilian development is the advent of commercial reentry capsules carrying micro-gravity experiments. Boosted aloft by space-launch vehicles, these reentry capsules may stay in orbit anywhere from a fractional orbital period to a couple of years before de-orbiting to return to Earth. One example is the German-Russian system to be available in 1999.[23] It is worth recalling that the SALT II Treaty banned fractional orbital bombardment systems for the United States and the former Soviet Union. Like reusable launch vehicles, these systems too may raise particular security concerns about the putative development of "space-strike" weapons. Dual use of reentry capsules and reusable launch vehicles may demand an international inspectorate for the maintenance of international peace and security in outer space.

All of these technological developments have not gone on in a vacuum of either policy or doctrine. Some quarters of government speak increasingly about the need to exercise "space control."[24] In its simplest form, "space control" means three things:

- maintain an assured access to outer space;
- ensure a defender's ability to use outer space; and,
- negate an attacker's use of outer space.

A central debate follows whether these goals can be achieved without resort to the "aggressive use" of outer space requiring force application systems such as ASATs. Many of these policy statements refer to benign or "nonaggressive" improvements in an era that witnesses military use of civilian satellites. Several examples will illustrate this point.

---

22. L. David, "Military Space Plane Sized Up For Battle," *Space News* (28 July 1997), 3.
23. "German-Russian Team to Build Return Capsule," *Space News This Week* (22 September 1997), 2.
24. M. Walsh, "USAF Targets Control in Air, Space," *Defense News*, 12:16 (21–27 April 1997), 12, 26.

Developments of the next generation reusable launch vehicles, known as single-stage-to-orbit systems, will offer the ability to launch reduced payload masses without ever leaving the confinement of the lower forty-eight states of the continental United States. Anyone having seen newscasts of the hover test of the Delta Clipper vehicle will recognize the ability of these systems to operate as "elevators to the sky." The X-33 Venturestar system may similarly ascend directly only to glide a corkscrew trajectory down. An upper stage attached to the payload satellite could give the released satellite the necessary impulse to orbit the Earth without the launch vehicle flying along the same initial trajectory. Unlike the Space Shuttle, which completes orbits of the Earth, these new launch vehicles would be more survivable against the threats posed by Earth-based ASAT weapons. The improved operational readiness of these vehicles may also provide "launch on demand" services to replenish small but militarily significant satellites should the operational requirements demand launching of satellites held in reserve during stressed environments.[25] Further speculation is that lasers held in the internal payload bays of a squadron of these reusable launch vehicles could be launched on short notice in a "launch surge" to bypass space-based anti-ballistic missile limitations of an ABM Treaty at some point in the future. Having never completed a single orbit, a single-stage-to-orbit reusable launch vehicle carrying an internal spaceborne laser would not technically fulfil the definition of space-based as it has come to be defined by the international community.

One measure to increase the survivability of satellite systems against ASAT activity is to move from single heavy satellite platforms to a "swarm" of many small satellites networked in a system. This trend shows up in mobile communications satellites where Iridium may very well compete with FLTSATCOM missions for US Navy needs. The US National Reconnaissance Office (NRO) has recently announced studies to replace large platforms with a swarm of smaller satellites to fulfil monitoring requirements.[26] The miniaturization and spatial diversity of satellite functions are an illustration of efforts to ensure the ability to operate in a threat environment that includes ASAT weapons.

Advocates for the ASAT mission argue that the third element of space control may not be performed without the possession of dedicated ASAT weapons.[27] Opponents argue that exercising selective "shutter control" for remote sensing data, or "access control" on the next generation, digitally-switched, and regenerative repeater communication satellites provided by US companies, may help dispense with the need for ASAT missions in the future.

---

25. F. Ferster, "Industry Moves Toward Launch-on-Demand Capability," *Defense News*, 12:41 (13–19 October 1997), 88.
26. J.C. Anselmo and P.J. Klass, "NRO Embraces Sigint Smallsats," *Aviation Week & Space Technology*, 147:13 (29 September 1997), 35.
27. W.B. Scott, "'Space Control' Shifting to 'Space Superiority'," *Aviation Week & Space Technology* (10 March 1997), 57–58.

Exploiting the terrestrial vulnerabilities of satellite infrastructures to existing conventional weapons and tactics is yet another means to accomplish the same end without resorting to ASAT activity.[28] The inherent capability of land-based lasers, air-based lasers, and high velocity land-, sea-, and air-based ballistic missile defence systems will also diminish the need for dedicated ASAT systems. Those advocating the aggressive use of space often overlook how the rule of law embodied in verifiable arms control agreements may provide greater levels of national security through reduced levels of threat. Arms control agreements, such as the INF Treaty, can remove the threat so that defence against it becomes unnecessary. Select individuals within some branches of military service, however, have not lost sight of how equitable treatment in outer space may enhance rather than diminish a nation's national security.[29]

Why the "Non-Weaponization" of Outer Space Terminology?

Whatever domain humanity has claimed dominion over has been stained by the blood of war. Battles have been waged on land, on and under the sea, and in the air. The pristine nature of outer space has thus far escaped "weaponization." To the best knowledge of the international community, outer space is today, a "weapons-free" environment. On the eve of humanity's permanent presence in outer space there arises the opportunity to ensure that in this next ascent we do not repeat our past falls. The legal regime for outer space, unlike the laws of land, sea, or air before it, is a unique environment that can first establish the rule of law.

The CD has examined the outer space dimension through different lenses in the past, trying to dissociate civilian from military, stable from unstable, or aggressive from nonaggressive activities. The CD has also tried to define bans along functional lines, such as anti-satellite versus ballistic missile defence versus space-strike weapons. The problem of outer space, however, may be best examined from the vantage point of the "engagement scenario" dimension in a multilateral arms control matrix. Table I illustrates the multilateral arms control matrix and the four engagement scenarios between weapon-basing and target-basing options. The four engagement scenarios are: Earth-to-Earth, Earth-to-Space, Space-to-Space, and Space-to-Earth. Here, Earth-based refers to land-, sea-, and air-based weapons. Space-based refers to those objects that complete at least one full orbit of any celestial body including the Earth, or are on trajectories to or from any celestial body excluding the Earth-to-Earth interaction, or are stationed in outer space in any other manner.

---

28. Press Briefing by OMB Director Frank Raines, NSC Senior Director for Defense Policy and Arms Control Bob Bell, and Deputy Defense Secretary John Hamre On the Line Item Veto, White House, Washington, D.C., 14 October 1997 as available at http://www.whitehouse.gov.

29. G.I. Seffers, "Army Urges Satellite Treaty," *Defense News*, 12:24 (16–22 June 1997), 4, 83.

Earth-to-Earth engagements would include the flight of ballistic missiles or fractional orbit bombardment systems. Earth-to-Space engagements would include direct ascent anti-satellite systems, such as the F-15-launched Miniature Vehicle of the 1980s era. Space-to-Space engagements would include the Soviet co-orbital anti-satellite system that required up to two orbits to intercept its target. Space-to-Earth engagements would include space-based lasers attacking reentry vehicles transiting outer space, or aircraft in flight.

The weapons classes of the multilateral arms control matrix are identified as weapons of mass destruction, conventional weapons, and "exotic" weapons. Weapons of mass destruction include nuclear, biological, and chemical weapons. Conventional weapons are weapons that rely on chemical explosion or kinetic energy impacts to destroy targets. This generally requires close proximity. "Exotic" weapons, borrowing language from the ABM Treaty, are based on "other physical principles." This category includes weapons such as lasers, neutral particle beams, or high-powered microwave beams that can destroy targets from a remote, stand-off distance.

Weapons systems, like any entity, have a life-cycle. They begin with research and transition to development. Development typically includes the tasks of design, manufacture, assembly, and testing of the prototype system. From development, weapons systems enter deployment, which includes mass production, quality testing, installation or fielding, and upgrades or repairs and overhauls of the flight systems. At the end of their operational lives the weapons systems may be decommissioned and destroyed to be replaced by the next generation. Use of weapons systems may proceed from any point after a prototype assembly until the last deployed system is successfully destroyed. Transfers of weapon systems may proceed from joint research to the decommissioning phases of the life-cycle. Table I illustrates the multilateral arms control matrix for the development life-cycle state of weapons systems pertaining to outer space and the status of multilateral attempts to prohibit these classes of weapons. Similar matrices may be developed for other life-cycle states of a weapon's product cycle.

The sparseness of the multilateral arms control matrix should calm the fears of those who question the future role of arms control negotiators at the end of the Cold War. Similarly, the first Earth-to-Earth domain of the multilateral arms control matrix shall probably remain uncompleted until the international community has successfully negotiated complete and general disarmament. The Earth-to-Space domain will likewise remain uncompleted until ABM systems and ballistic missile systems can be eliminated through a total ballistic missile ban treaty (TBMBT). The remaining two columns denote the region in which a convention for the non-weaponization of outer space would apply. Such a convention would also apply to the development, testing, and use of life-cycle states of the space-based weapons product cycle. Non-transfer obligations would also be a part of the convention to ensure that no nation benefits by remaining outside of it.

Table I. Multilateral Arms Control Matrix

Weapon's DEPLOYMENT Life-Cycle State	**ENGAGEMENT SCENARIO (Weapon-Basing to Target-Basing)**			
WEAPON CLASS	**Earth -to- Earth**	**Earth -to- Space**	**Space -to- Space**	**Space -to- Earth**
WMD	a,c	b,c	X c,d	X c,d
Conventional	e	e	d	d
"Exotic"	f	f	d	d

Key:

X (shaded)	Complete prohibition of weapon class with substantial multilateral participation.
(shaded)	Partial prohibition, or restriction(s) on weapon class with substantial multilateral participation.
(unshaded)	Unfettered freedom of action for a substantial population of the international community.

a Partial prohibition as CWC and BTWC only agreed to thus far. Nuclear element of WMD is still outstanding for weapons that "transits through" outer space.

b Partial prohibition as CWC and BTWC only agreed to thus far to prohibit deployment of WMD weapons that "reaches into" outer space. Nuclear component is still outstanding.

c The OST, LTBT, CTBT, CWC, BTWC collectively prohibit deployment of WMD that are space-based. Space-based means objects that complete at least one full orbit of any celestial body including the Earth, or on trajectories to or from any celestial body excluding the Earth-to-Earth interaction, or stationed in outer space in any other manner.

d Potential entries as a result of a convention for the non-weaponization of outer space. Prohibition would also apply to the development, testing, and use life-cycle states of a weapon's product cycle.

e Potential entries as a result of a total ballistic missile ban treaty for weapons that "transits through" or "reaches into" outer space. Prohibition would also apply to the development, testing, and use life-cycle states of a weapon's product cycle.

f Potential agreement restricting Earth-based "exotic" weapons based on "other physical principles" such as lasers that "reaches into" outer space. It might not be possible to differentiate between lasers capable of an anti-armour role and those capable of an anti-satellite role.

A convention for the non-weaponization of outer space would not try to roll back the clock and remove the military presence from outer space and would accept the current "nonaggressive" military use of space that contributes so much to international peace and security. In this vein, such a convention is not analogous to the Antarctic Treaty insofar as that it does not require that outer space be used for "exclusively peaceful purposes." Instead, the convention would be a pragmatic and realistic set of forward looking constrains. Similar to the aim of the Non-Proliferation Treaty before it, a convention for the non-weaponization of outer space would stop the proliferation of space-based weapons. It would essentially cap the migration of missions performed by Earth-based weapons systems into missions performed by space-based weapons systems before the operational existence of these putative weapons systems. The convention would further protect from the continuing march of conventional and exotic technology so necessary for continued international peace and security in a millennium that will ultimately see humanity establish a permanent presence on some other celestial bodies in the solar system. Successful negotiation of such an act would essentially complete the lower right-hand corner of the multilateral arms control matrix. It is well recognized that it is more cost-effective for the international community to prohibit weapons systems through non-proliferation or arms control agreements than to wait until the much tougher negotiations must begin on disarmament agreements.

A convention for the non-weaponization of outer space would not initially attempt to encompass existing Earth-based systems that "reach into" outer space or prohibit those currently planned systems necessary to counter the perceived proliferation threats posed by ballistic missiles armed with weapons of mass destruction. This is a hard political choice. Nevertheless, the lessons learned from successful negotiation of such a convention would have direct applications to the subsequent negotiation of, for example, a TBMBT. Having first removed the threat of space-based weapons, the convention would enable negotiations on a TBMBT to eliminate those technologies that could serve as anti-satellite systems. As a result, the convention is but one step toward the goal of complete and general disarmament that will ultimately see the transition of military forces to police forces. At that end state of arms control initiatives, outer space shall truly be used for "exclusively peaceful purposes."

Negotiation of a convention for the non-weaponization of outer space has some prospects for success in the current political environment. The recent demarcation agreements between the State Parties of the Memorandum of Understanding (MOU) Regarding the ABM Treaty and the willingness on the part of Russia and the United States to address the anti-satellite problem may very well signal the desire for the prevention of the space-basing of weapons. Conference on Disarmament activity in this direction would further help to establish the spirit of a newly agreed international norm from which other elements of the multilateral arms control matrix could then be negotiated.

Could such a Convention be Verifiable?

Recent American BMD policy — as enunciated in companion material to the MOU Regarding the ABM Treaty — recounted how difficult it would be to differentiate space-based weaponry for TMD purposes from space-based weaponry for ABM purposes in justifying further limitations on space-based weapons.[30] Such disclosures illustrate how a complete space-based weapons ban from outer space might be easier to verify than those conventions which seek prohibitions based on functional definitions. By extension then, one might envisage how space-based ASAT prohibitions could also be promoted with a similar argument. What is certain from this musing, however, is that a convention that bans all space-based weapons should be easier to verify than those conventions that ban only some weapons from outer space.

Verification of a convention for the non-weaponization of outer space will also depend on which stages of the product life-cycle the convention shall specifically encompass. Many arguments have been made to the effect that a convention which prohibits testing and deployment should be easier to verify than those that will ban development, testing, deployment, use, and destruction.[31] Canadian studies reinforce others which show that on-site inspections, terrestrial-based space surveillance,[32] and space-based in-orbit monitoring,[33] can verify a suitably framed space-based weapons ban. The former Soviet Union, through Chernenko, is on record that "Effective verification of compliance by the Parties with a moratorium on orbital-effect anti-satellite weapons could be assured by the means for the tracking of space objects which the Parties have at their disposal."[34]

Early warning and space surveillance technologies are considerably improved in their own right since 1984. Deployment of next-generation sensors, such as the US SBIR-High and SBIR-Low systems, will provide an unmatched capability to track objects ascending through the atmosphere and operating in outer space which is necessary to perform the ballistic missile defence mission.[35] Hints at terrestrial-based capabilities occurred with the

---

30. United States Congress, Office of Technology Assessment, "Anti-Satellite Weapons, Countermeasures, and Arms Control," OTA-ISC-281 (Washington: US Government Printing Office, September 1985).
31. *Ibid.*
32. C.A. Rutkowski, *The Role of Astronomical Instruments in Arms Control Verification*, Arms Control Verification Studies No. 2 (Ottawa: External Affairs Canada, 1986).
33. *The Paxsat Concept*, Verification Brochure No. 2 (Ottawa: External Affairs Canada, 1986).
34. Letter dated 16 June 1984 from the Representative of the Union of Soviet Socialist Republics addressed to the President of the Conference on Disarmament transmitting the text of the answers by Mr. Konstantin Chernenko, General Secretary of the CPSU Central Committee and Chairman of the Presidium of the Supreme Soviet of the USSR, to questions by a United States Journalist Mr. Joseph Kingsbury-Smith, CD/510, 18 June 1984.
35. W. Ferster, "Satellite System Could Track Threats," *Defense News*, 12:23 (9–15 June

first space shuttle flight when ground-based national technical means were used to observe the missing tiles from the surface of the space shuttle. These terrestrial-based observation systems for outer space have also seen considerable improvements since 1982. Here, systems like the US Advanced Electro-Optic System (AEOS) will provide the ability to recognize every object launched into low Earth orbit. Adaptive mirrors, laser atmospheric turbulence compensation, and advanced signal processing techniques displayed in civilian astronomy can also help in the verification of a space-based weapons ban. The key to any such verification of outer space will be through cooperative monitoring.

Cooperative monitoring could be a breakthrough verification technique that would share only necessary and sufficient data collected by national technical means to verify an outer space convention at minimal cost. The precedence for this form of verification had been established with the series of seismic monitoring experiments preceding the establishment of the CTBT verification system. It has also been reported that on several occasions since 1992, the SpaceTrack catalogue of NORAD had been exchanged with the Russians.[36] US Presidents Reagan and Bush also offered to share data and technology with the former Soviet Union/Russia for ballistic missile defence purposes. Current US President Clinton has authorized a broader distribution of early-warning data from NORAD to include NATO, Japan, and Israel.[37] The United States also now shares classified imagery with partners in Bosnia in support of international peacekeeping.[38] Such cooperative undertakings by nations would further reinforce the *res communal* basis of outer space and forestall the costly requirement to build numerous national verification networks.

In terms of on-site inspection, many conventions have been signed by the current signatories of the Outer Space Treaty to authorize such undertakings by foreign nationals with the result that national legislative assemblies are now familiar with the requirements for legislation and regulations necessary to permit such activity in their respective nations. In 1967 such was not the case. It is also worth recalling that the Outer Space Treaty already contains a provision for on-site inspections. Article X reads:

1997), 3.

36. Col. Nikolai Belov, Alexander Chebotarev Missile and Space Defense Forces HQ, "Attention: Garbage in Outer Space," found at http://www.milparade.ru/17/139-141.htm.

37. "Sustaining the Strategic Space Advantage," prepared statement of Air Force General Howell M. Estes III, Commander-in-Chief, North American Aerospace Defense Command and U.S. Space Command, testimony before the Senate Armed Services Committee, 13 March 1997.

38. Honourable Keith R. Hall, Director NRO, Speech to National Space Club, Washington, D.C., 15 September 1997.

> In order to promote international co-operation in the exploration and use of outer space ... State Parties to the Treaty shall consider on the basis of equality any requests by other State Parties to the Treaty to be afforded an opportunity to observe the flight of space objects launched by those States. ... The nature of such an opportunity for observation and the conditions under which it could be afforded shall be determined by agreement between the States concerned.

As far as is known, not a single nation has exercised such privileges for verification of the Outer Space Treaty. Managed access provisions would assist verification efforts for application in the development life-cycle state.

To overcome verification difficulties relating to use of space-based weapons, suitably designed Situation Observation and Reporting System (SOARS) piggy-back payloads, analogous to airborne flight recorders or "black boxes," could be flown on satellites at the discretion of the satellite operators. These payloads were developed in the US after purported illumination of US reconnaissance satellites by Soviet ground-based lasers.

Verification, equity, and balancing international and national security interests will all play an important role in developing a convention for the non-weaponization of outer space. "The Devil is," as they say, "in the details." What is truly important now is that we begin this important undertaking before technological and political events overtake the ability of the Conference on Disarmament to act.

A Proposed Approach Within the Conference on Disarmament

Encouraged by the approval of UN/RES/59/37, Canada has suggested the early appointment of a Special Coordinator to explore with CD members an approach along the above lines to satisfy the international community's desire to preserve outer space for peaceful purposes. Subsequent to this appointment, an Ad Hoc Committee should be formed with a mandate to prepare the necessary groundwork in terms of scope, definition of technical and legal parameters, and evaluation of the means by which the convention for the non-weaponization of outer space could be verified and administered in preparation for subsequent negotiation of a convention. This preparatory phase should be of a limited duration. An Ad Hoc Committee with the mandate to negotiate such a convention should be the natural outcome of this process. This negotiation phase should be consistent with a goal of obtaining a convention by 2001.

Other means for attaining the goals of such a Convention, including a review of the Outer Space Treaty along the lines of the Bacteriological and Toxins Weapons Convention experience, or a replay of the Ottawa Process for a Landmines Treaty, are available to a sufficiently motivated international community to ensure that no nation mistakes the Conference on Disarmament rules of consensus as *de facto* expressions of national vetoes. Economic penalties, like those proffered in the Chemical Weapons Convention (CWC) for States Parties that choose to remain outside of that Convention, as well as economic incentives, such as the distribution of royalties from extraterrestrial resources to those States which choose to be Parties to the convention, may provide additional incentives for obtaining wide participation in a convention for the non-weaponization of outer space.

Conclusion

Arms control agreements must be measured by three yard sticks: they must be equitable; they must be verifiable; and they must be within the national security interests of the nation considering signing the agreements.

A convention for the non-weaponization of outer space would disadvantage no single nation more than any other before the operational existence of space-based weapons. In this sense, banning weapons from deployment in outer space is an equitable agreement.

The appropriate technology, processes, and procedures from prior arms control agreements could be adopted to ensure that such a convention would be verifiable. The second requirement for arms control agreements is thus attainable.

By removing the direct threat that space-based weapons would pose to nations and by ensuring unimpeded progress in nuclear disarmament and non-proliferation goals, a convention for the non-weaponization of outer space is in the national security interests of every nation.

Chapter Seven

Confidence Building and Outer Space

James Macintosh*

This chapter provides a very brief introduction to confidence building and offers several observations about the potential use of confidence building in the context of outer space. The appendices include a summary table of generic confidence building measures (CBMs), a checklist of CBMs for possible use in the outer space context, a summary table of the underlying assumptions animating various practical understandings of confidence building, and a brief discussion of non-traditional and unconventional security concerns. The main intent is to provide those interested in international law and arms control in space with some food for thought relating to the use of the confidence building approach.

The Conventional Concept of Confidence Building

Confidence building usually is thought of as the use of measures designed to improve information, increase understanding, and reduce uncertainty about neighbours' military forces and activities (including both capabilities and intentions). An improvement in security relations (or, possibly, even broader relations) is thought to occur as a result of the increased availability of information (made available on a cooperative basis) and the increased opportunities for interaction among personnel and officials from the security communities of participating states. Modest constraints on forces and their activities can also play a role, although not everyone accepts this view. Most understandings of confidence building assume that the states pursuing the approach have difficult security relationships marked by concerns about the military intentions and capabilities of their neighbours. Confidence building offers a way of moderating these difficult and potentially dangerous security relations. This generally is seen to be the principal motivation for engaging in confidence building.

---

* This chapter is an edited version of a paper commissioned for the Fifteenth Annual Ottawa NACD Verification Symposium, "Arms Control and the Rule of Law: A Framework for Peace and Security in Outer Space," 11–14 March 1998, Montebello, Quebec, Canada. The views presented in this chapter are those of the author and do not necessarily represent those of the Department of Foreign Affairs and International Trade or the Government of Canada.

Conventional understandings also tend to be based on efforts to deal with traditional security threats such as those posed by relatively large military forces (usually land forces) in close proximity to each other. This is an obvious legacy of the European conventional force security environment of the Cold War era, the original context in which most confidence building ideas were developed. Thus, contemporary understandings of confidence building are typically derived (whether explicitly or implicitly) from the practical historical experience of the CSCE (now termed the OSCE) and its comprehensive confidence building agreements (principally the various Vienna Documents). There is also a relatively distinct body of confidence building thinking associated with the maritime security application context. In this case, "Incidents at Sea" type arrangements (activity constraint) are typical of confidence building efforts, although information, interaction, monitoring, and other constraint measures also have been applied or explored.

Recently, some attention also has been devoted to exploring how confidence building can assist in moderating non-traditional security concerns such as drug trafficking, piracy, international terrorism, undocumented migration, and environmental and resource management concerns. For some, these non-traditional security concerns represent a more relevant threat to fundamental security and thus a more appropriate focus for "confidence building" efforts. Conventional accounts of confidence building, however, have some trouble dealing with these sorts of issues because many standard CBMs constitute a relatively poor fit for managing these types of problems. Worse, most conventional accounts assume difficult security relationships among participants and this may not be the case with some non-traditional security concerns. This is potentially relevant to the outer space application area, as we will see later. Very little attention in the professional literature has been devoted to unconventional security concerns.[1]

The traditional literature and application efforts based (explicitly or implicitly) on this literature are marked by considerable conceptual imprecision. These traditional understandings tend to be:

- Measure-centric (confidence building *is* the use of standard CBMs);
- Distinctly operational (focusing on what types of CBMs to use and how to apply them);
- Conventional force-oriented;
- Conceptually unsophisticated (lacking any clear sense of causal explanation); and,

1. These contexts are discussed briefly in James Macintosh, "Diverse Interpretations of Confidence Building," a paper prepared for Canada's Department of Foreign Affairs and International Trade (November 1997). It also appears in *Non-Proliferation, Arms Control and Disarmament: Enhancing Existing Regimes and Exploring New Dimensions*, ed. by Peter Gizewski (Toronto: Centre for International and Security Studies, York University, 1998). Appendix IV contains a summary of this material.

- Silent about necessary or sufficient supporting conditions.

Thus, the traditional literature fails to provide a plausible account of how and under what conditions confidence building can produce or contribute to a meaningful improvement in security relations. It also tends not to explore in much depth the broader or deeper lessons of successful application examples, particularly in the OSCE context. Finally, it encourages policy-makers to think that applying one or more CBM-like measures (either in a separate agreement or as part of some broader type of arrangement) counts as confidence building. As a result of these typical shortcomings, a number of practical understandings of what confidence building means have proliferated in both the policy world and in the work of experts. These practical understandings can have diverse and contradictory underlying assumptions making them incompatible with each other. Appendix III contains a summary of these understandings and their assumptions.

These shortcomings are very limiting and cannot help but impair the successful use of the confidence building approach in new or extended application contexts. They can also lead to the premature pursuit of the approach.

A Revised Understanding of Confidence Building

In order to move constructively in the direction of more effective policy, it may prove helpful to develop a more general and flexible, yet rigorous, understanding of confidence building, one not tied so operationally to existing examples of CBMs and dismissive of process and causal issues. Such a collaborative endeavour, hopefully involving experts and policy-makers from many existing or potential application contexts, need not reject outright the central, common elements of the traditional approach with its measure-centric focus on transparency, interaction, and (sometimes) constraint. Instead, it can supplement it with a clearer sense of how the confidence building process works and under what conditions.

One such revised understanding (the *transformation view*)[2] argues that the confidence building *process* is effective because its fundamental nature is well suited to helping *restructure security relations, moving them in more cooperative directions.* The transformation view sees confidence building as being more than CBMs and what they do or require. Instead, confidence building is seen to be an activity that, because of its basic nature, can structure and facilitate constructive change in security relationships. It may very well be extendable to non-traditional security concerns and unconventional security issues as well.

This understanding recognizes that certain key supporting conditions must be present in a security relationship in order for a real process of change

---

2. See James Macintosh, *Confidence Building and the Arms Control Process: A Transformation View,* Arms Control and Disarmament Studies No. 2 (Ottawa: Department of Foreign Affairs and International Trade, 1996).

to be possible. New, more cooperative rules, practices, and principals of security behaviour emerge from this process (including but not limited to the specific requirements of CBMs), helping to define the nature of the revised security relationship.

According to this revised understanding, confidence building is a distinctive activity entailing the three associated processes of exploring, negotiating, and implementing generally cooperative measures dealing with specific security concerns. In the case of traditional military concerns, they typically include interaction, information, and constraint measures, but other types of measures are also possible. The measures (CBMs) *and* the behavioural practices and principles associated with their development and implementation embody the key elements of the more cooperative relationship. When conditions are supportive, this process can facilitate, focus, amplify, and structure the potential for a positive change in relations.

Thus, the confidence building process, by its nature, structures and synchronizes a developing desire to revise strained and unsatisfactory relations that no longer seem to work satisfactorily according to the "old (prevailing) rules." The confidence building process provides a distinctive framework for the at least partially cooperative development of revised relationships structured at least in part by new, more cooperative rules and practices. While competitive elements will almost surely remain in the relationship — most states, after all, will continue to have disputes about a variety of important issues — its basic character will shift significantly in a cooperative direction.

This revised understanding of confidence building assigns supporting conditions an important role. In the case of traditional security relations, unless a security relationship is characterized by most of these emerging conditions, confidence building will not seem attractive and likely will not be successful if attempted. In the main case of traditional security relations, the supporting conditions appear to be:

- "Security management fatigue";
- Unease and dissatisfaction with *status quo* security policies;
- Concern about various costs associated with maintaining the *status quo*;
- The existence of an experts group (preferably with cross-national membership or ties) to support and explain confidence building;
- A new generation of policy-makers with broader international experience and a willingness to consider new ideas;
- Forum(s) for bilateral or multilateral discussion (formal and/or informal); and,
- A "leap-of-faith" or context-altering initiative by at least one major participant.

A key factor in developing successful confidence building arrangements is the need to recognize that participants must formulate appropriate measures themselves and do so with conviction. This is an important and integral aspect of the confidence building process. Neatly packaged CBM "solutions" imported by outsiders will likely short-circuit this important dimension of confidence building. An imperfect understanding of how the confidence building process works successfully may not prevent success but it cannot facilitate it and may destroy or undermine worthwhile opportunities. An imperfect understanding may also encourage confidence building efforts when conditions cannot support them.

The point in developing this alternative or supplementary understanding of confidence building is not simply to add one more version of what confidence building is to the existing list of practical understandings. This reconstructed understanding is intended to explain, at least in a preliminary fashion, why it is that pursuing confidence building can produce a constructive outcome and under what circumstances it can do so. At the very least, it identifies the sorts of questions that we ought to be asking when we explore the nature of confidence building and its potential applicability in new contexts.

Confidence Building and Outer Space

How we view the opportunities for confidence building in the outer space application context will depend upon how we think about confidence building more generally. If our thinking is informed by the traditional or conventional understanding, then it will seem very natural to use a comprehensive list of existing CBMs as our main reference. After all, according to this view, confidence building as a distinct activity is more-or-less defined by the content of standard CBMs — confidence building *is* the use of CBMs.

Exploring confidence building opportunities then becomes a relatively straightforward task entailing the matching of CBM types with outer space "problems." An outline of what this might entail is included in the appendices. Thus, in perhaps oversimplified terms, confidence building in this application context amounts to the development and use of standard CBMs (and their variants) in order to address security concerns directly or indirectly resulting from activities and capabilities in outer space.

Typically, we would expect to see politically binding agreements and legally binding treaties emerge from this endeavour, although less formal arrangements could also emerge. There likely also will be a tendency to see confidence building products as less demanding than arms control products, paralleling the experience of the CSCE case which inevitably informs most thinking (either implicitly or explicitly) about these approaches and their relationship to each other. There is a strong inclination to view confidence building as something to explore first because it is easier and less-demanding in virtually all respects.

It is an open question whether discussions of confidence building in the outer space context, based on the traditional understanding, might attempt to reinterpret past international efforts (The Outer Space Treaty, The Moon Treaty, and the ABM Treaty, for instance) in terms of whether or not they exhibited characteristics of a confidence building negotiation. In principle, such insights might provide guidance for current or future attempts. It seems far more likely, however, that confidence building explorations will seek to apply existing CBM ideas to the solution of current security problems associated with outer space. "New ground" in this context is likely to be limited to unusual applications of existing CBMs or, possibly, the *ad hoc* development of unconventional CBMs suited to the special security problems associated with activities and capabilities in outer space. In short, it is likely to be measure-centric, pragmatic, and future-oriented.

Although the transformation view also takes the existing constellation of CBMs seriously as a basis for thinking about confidence building, it focuses on other important concerns, as well. For instance, this understanding of confidence building also asks whether participants are dissatisfied with the existing structure of security relations (generally or in a more limited sense restricted to some subset of security relations or issues) and focuses on changes in practice and principle that can help constitute a revised (typically more cooperative) structure of security relations. In its fullest form, it asks what fundamental changes in agent (policy-maker) thinking might occur as part of the confidence building process and how this could be manifested in structural and procedural change (new institutions or regimes) relevant to the security concerns in question.

Given this perspective, it is natural to ask whether historical negotiations in the outer space context have, in fact, exhibited characteristics of confidence building. The focus shifts from measures to constructive changes in security relationships (represented by the institutionalization of new practices and principles) resulting from the development of cooperative measures that may or may not share common features with existing CBM categories. It is at least possible to argue that the processes that produced (especially) The Outer Space Treaty, as well as The Moon Treaty, exhibited some characteristics that seem similar to what the transformation view identifies as important. This is a possible insight that is unlikely to emerge from the use of more traditional understandings. Understanding these transformation view-related insights may prove very helpful in developing future efforts.

In the arguably more important case of future application, the transformation view encourages us to ask similar questions. These questions include:

- Is there substantial dissatisfaction with the *status quo* in the realm of the outer space-related security policies of key states?
- On whose part and why?

- What are the concerns and can more cooperative (or even collaborative) practices and principles address these concerns?
- What practices and principles are there that might accomplish this?
- What specific measures might help animate this revised structure of security relations?
- What form of exploratory and negotiating forum(s), featuring what fundamental characteristics (i.e., generally competitive or cooperative), would be best suited to addressing the relevant security concerns and their possible solutions?

While it is relatively clear that the military use of space[3] can be identified as a major concern of many states, either in the context of ballistic missile defence or information gathering and military support, it is not clear in what way(s) confidence building could help address these concerns in the near-term future. On first blush, it seems more likely that zero-sum thinking more akin to traditional arms control practices will be more likely to address these concerns effectively (if they can be addressed at all). However, this assessment overlooks the degree to which the superpowers were willing to embrace surprisingly far-reaching and cooperative understandings in the negotiation of The Outer Space Treaty and The Moon Treaty.[4] The same may be true of the ABM Treaty. If this interpretation is correct, it demonstrates that a major restructuring in the direction of cooperative principles regarding space policy by keenly involved participants is possible. Perhaps a similar type of shift in basic thinking can be facilitated in the context of contemporary issues.

However, as we confront the difficult problems of space weaponization associated with, for instance, unilateral ballistic missile defence, it does seem more likely that arms control limitation agreements (for instance, a revised ABM Treaty) might offer the prospect of a more constructive and effective resolution. Conventional confidence building thinking does not seem to provide many workable ideas for addressing in a central way these sorts of issues. The best that it can offer seems to be limited to PAXSAT-style monitoring and inspection of space objects, technology constraints (specifically, bans on testing certain types of systems), and deployment constraints.

One confidence building-related alternative that might prove both attractive and relevant given this assessment is the notion of "confidence expand-

---

3. This includes: placement of offensive or defensive weapons (and their support systems) in space for use in space or against terrestrial targets; placement in space of systems to directly facilitate terrestrial weapons use; placement of monitoring and surveillance systems with clear military applicability in space; and, the development and deployment of surface- or air-based weapons that destroy objects in space.

4. For an excellent discussion of this, see M. J. Peterson, "The Use of Analogies in Developing Outer Space Law," *International Organization*, 51:2 (Spring 1997). Peterson uses a cognitive science approach to explain how "law of the sea" and "Antarctic Treaty" ideas informed thinking about space and celestial bodies.

ing."[5] This major extension of confidence building thinking, developed in the context of the Caribbean island states, envisions the cooperative development of collaborative approaches for the effective management of non-traditional security concerns. The main idea here is that collaborative efforts, where security resources are pooled and jointly administered, can (among other things) offer the capability (and sometimes the *only* capability) to successfully address security threats originating from outside the group of participating states *without* at the same time posing a threat to any of the participants. Multilateralizing the provision of defence against security threats from space through the use of space-based assets (aimed principally against the use of missiles by "rogue" or "crazy" states as well as unauthorized missile launches) could offer a way to withdraw from the thicket of difficult issues currently associated with intrinsically unilateral space-based defences. This is a possibility that warrants further exploration as an alternative to unilateral action or arms control limitation negotiations.

Conclusion

The confidence building approach has established a significant, if poorly understood, record of accomplishment in the last twenty or so years. It is a potentially important and valuable security management approach that can help restructure security relations — and, potentially, other types of international relations — in what are likely limited but important ways *if* the participating states are ready for change because they are dissatisfied with *status quo* policies. However, positive change is neither automatic nor inevitable and is unlikely unless participants themselves are ready for change in their security thinking and broader relations. Simply attempting to apply existing categories of CBMs to new application contexts is unlikely to yield constructive results. It is only by thinking more carefully and deeply about what it is that makes confidence building function effectively that we can hope to use this approach in a constructive and useful manner.

Applying new insights about confidence building to the outer space application context, particularly those that can be derived from understandings like the transformation view, may suggest that there are constructive options available for restructuring *status quo* approaches to the military use of space. In particular, elaborating on the confidence expanding idea may permit us to envision a dramatically new way of thinking about some of these very difficult issues, such as the unilateral pursuit of space-based defences that confronts us today. For this to happen, however, the potential participants of a restructured security regime in space will need to recognize that current

---

5. See James Macintosh and Ivelaw L. Griffith, *Confidence Building and Related Approaches: Managing Caribbean Security Concerns* (Ottawa: Department of Foreign Affairs and International Trade, 1996).

policies (and the thinking underlying them) are not effective and pose the risk of producing more harm than good. It will also require an unconventional, thoughtful, and well-conceived initiative that can address these nascent feelings by suggesting a powerful alternative. We may well be on the cusp of dramatic developments in the realm of space policy. If so, it requires an actor with vision, initiative, patience, and a deep understanding of the issues involved to take the first, bold step.

APPENDIX I — CBM TYPOLOGY[6]

An effective way of representing what confidence building encompasses operationally is the construction of a comprehensive collection of CBM categories, defined by basic function. The general categories listed below are of "traditional" CBMs, ones that have been suggested for or used in addressing concerns about conventional military activities and capabilities. Many of these are also suitable (with appropriate adjustments) for dealing with non-traditional security concerns. It is quite likely that further efforts developing confidence building for use in addressing non-traditional security issues will suggest fundamentally new types of CBMs that, in turn, might prove useful in more traditional applications. Variants of these traditional CBMs also might prove useful in confidence building efforts intended to deal with unconventional security threats. There has not been much work done on new CBMs specifically designed to address unconventional security concerns. This is an area that requires further research.

TYPE A: INFORMATION-ORIENTED CBMs[7]

(1) **Information Measures** (provision of information about military forces, facilities, structures, and activities)
Examples include: publication and/or exchange of weapon system and force structure information; consultative commissions; publication and/or exchange of defence budget figures; and, publication and/or exchange of weapon system development information.

(2) **Interaction Measures** (provision of opportunity to interact with opposite numbers)
Examples include: military personnel exchanges; security expert exchanges; transnational secondments; joint training and exercises; and, seminars discussing doctrine, strategy, and technology issues.

(3) **Communication Measures** (provision of means of communication)
Examples include: hot lines for exchange of crisis information; joint crisis control centres; and, "cool lines" for the regular distribution of required and requested information.

(4) **Notification Measures** (provision of advance notification of specified military activities)
Examples include: advance notification of exercises, force movements, and mobilizations — including associated information about forces involved and their equipment.

TYPE B: VERIFICATION AND MONITORING CBMs[8]

(1) **Observation-of-Movement Conduct Measures** (provision of opportunity to observe specified military activities)

---

6. This typology is a further development of one contained in *Confidence Building and the Arms Control Process: A Transformation View*. Note that, in practice, CBMs can combine elements of several categories. These are "compound CBMs."

7. Negotiations that focus on "transparency" usually contain only these types of measures.

8. This category has been — and remains — problematic. Although clear *monitoring*-type measures appear to be non-controversial members of this group, the status of

Examples include: mandatory and optional invitations to observe specified activities (with information about the activity) and rules of conduct for observers and hosts.

(2) **General Observation Measures** (provision of opportunities to engage in non-focused "looks" at relatively small and generally-specified sections of territory — not to be confused with inspection measures)
Examples include: Open Skies-type arrangements and roving (ground-based) observation teams with a general mandate to "look around for troubling activities."

(3) **Inspection Measures** (provision of opportunity to conduct detailed inspections of constrained or limited military forces, facilities, structures, and activities)
Examples include: generally limited to on-site inspections of various forms.

(4) **Monitoring Measures** (provision of opportunity to monitor — usually on an ongoing basis — constrained or limited military forces, facilities, structures, and activities)[9]

TYPE C: CONSTRAINT CBMs

(1) **Activity Constraint Measures** (provision of assurance to avoid or limit provocative military activities)
Examples include: no harassing activities such as "playing chicken" on the high seas or near territorial boundaries; no large-scale war games or maneuvers near territorial boundaries; and, no threatening equipment tests near sensitive areas.

(2) **Deployment Constraint Measures** (provision of assurance to avoid or limit the provocative stationing or positioning of military forces and equipment[10]
Examples include: no threatening deployments near sensitive areas (such as tanks on a border); equipment constraints such as no attack aircraft within range of a neighbour's rear area territory; manpower limits; and, nuclear-free zones.

(3) **Technology Constraint Measures** (provision of assurance to avoid or limit the development and/or deployment of specified new military technologies, includ-

---

specific *inspection* measures with clear verification intent is troubling. Should they be considered CBMs or associated verification measures? Perhaps the best way of dealing with this problem is to consider them CBMs so long as they are an integral part of a confidence building agreement but also recognize that they serve an associated purpose as distinct verification measures. The issue of whether integral verification measures should be considered CBMs requires further study, although it does not appear to have a material impact on the practical application of confidence building ideas.

9. The distinction between "inspection" and "monitoring" can be confusing. Inspection generally involves examining a specific item or activity, often after-the-fact, in order to determine compliance with specific undertakings. Monitoring generally involves a less-focused, ongoing process of watching. "General observation" is broad while "monitoring" is more focused. The differences are best understood by comparing: an on-site inspection to determine whether certain items of equipment have been destroyed; the use of a monitoring device to detect or record the removal of equipment from an enclosed storage area; and, an "open skies" type flight to examine whether any new facilities have been constructed in a designated area.

10. This type of measure is concerned with the presence and/or number of certain types of equipment or forces. Technology constraint measures focus on specific characteristics of equipment and the limitation of their improvement.

ing systems and subsystems, believed by participating states to have a destabilizing character or impact)
Examples include: no *replacement* of deployed military equipment of certain types (typically, tanks, heavily armored combat vehicles, self-propelled artillery, combat aircraft, and combat helicopters) with new, more advanced types; no *modernization* of deployed military equipment of certain types in certain key, well-defined respects; no *training* with new systems; no *field testing* of new designs; and, no *production* of specified new systems or subsystems.

APPENDIX II – CBM OPTIONS

	CATEGORY OF THREAT					
CBM TYPE	WEAPONIZATION (TERRESTRIAL)	WEAPONIZATION (SPACE AND TRANS-SPACE)	MILITARY UTILIZATION AND COMBAT SUPPORT	MILITARY INFORMATION COLLECTION	NON-MILITARY INFORMATION COLLECTION	CARELESS USE OF SPACE RESOURCES
INFORMATION ORIENTED						
INFORMATION						
INTERACTION						
COMMUNICATION						
NOTIFICATION						
VERIFICATION AND MONITORING						
OBSERVATION OF MOVEMENT CONDUCT						
GENERAL OBSERVATION						
INSPECTION						
MONITORING						
CONSTRAINT MEASURES						
ACTIVITY						
DEPLOYMENT						
TECHNOLOGY						

APPENDIX II – TRANSFORMATION VIEW-RELATED ADDITIONAL CONSIDERATIONS

	CATEGORY OF THREAT					
REGIME INDICATOR	WEAPONIZATION (TERRESTRIAL)	WEAPONIZATION (SPACE AND TRANS-SPACE)	MILITARY UTILIZATION AND COMBAT SUPPORT	MILITARY INFORMATION COLLECTION	NON-MILITARY INFORMATION COLLECTION	CARELESS USE OF SPACE RESOURCES
DEGREE OF COOPERATION - EXISTING (H, M, L)						
DEGREE OF COOPERATION - POTENTIAL (H, M, L)						
EXISTING BEHAVIOURIAL RULES						
EXISTING AGREEMENTS OR TREATIES						
EXISTING PRINCIPLES						
EPISTEMIC COMMUNITY SUPPORT						
SIGNIFICANT STATE OPPONENTS						
SIGNIFICANT STATE SUPPORTERS						

APPENDIX III – CONFIDENCE BUILDING VARIANTS AND COMPONENT ASSUMPTIONS

CONFIDENCE BUILDING VARIANT	BASIC OBJECTIVE MOTIVATING INTEREST IN CONFIDENCE BUILDING	NEGOTIATING ASSUMPTIONS	PREFERRED OUTCOME	UNDERLYING CAUSAL MODEL
IMPOSED	Contain the threat posed by target state; perhaps encourage rehabilitation of target state. Target state objective likely is to minimize impact of measures.	Not applicable. Measures are imposed. Some highly competitive negotiation of application requirements around the margins.	Continuing constraint of target state (largely through imposed transparency and capability constraints) and (possibly) rehabilitation of target state.	Enforced transparency and constraint (limits) will prevent or limit target state bad behaviour in future. Target state may "learn" to behave better.
COMPETITIVE CONSTRAINT	Constrain surprise attack or coercive capabilities of threatening neighbour while retaining maximum flexibility and capability.	Highly competitive; very suspicious; starkly zero-sum. Valued tradeoff likely price for some constraint.	Reduced or (less likely) eliminated opportunity for surprise attack or coercive use of force. Possibly reduced defence costs.	Skillful bargaining can bring disproportionate advantages. Tradeoff may reduce threatening capabilities or activities.
BASIC	Minimize chances of increased tension or unintended conflict due to misperception and overreaction. Perhaps also constrain surprise attack possibilities.	Competitive; suspicious; moderated zero-sum thinking with some emergent cooperative elements (as process progresses).	Reduced uncertainty; tensions; and chance of unintended conflict. Possibly reduced defence costs.	Increased information and interaction opportunities in modestly cooperative environment will reduce suspicion and tension. Modest constraints can help, too.
ENHANCED	Reduce level of tension and suspicion by developing new, more cooperative practices and principals for some troublesome aspects of security relations ("normal" form). Possibly, develop fundamentally restructured security relations ("advanced" form).	Mixture of competitive and cooperative-style thinking and negotiating behaviour with elements of cooperative style of thinking emerging as increasingly important and appropriate.	Revised security relationship with new, more cooperative practices and principles replacing some elements of status quo security policies. Reduced defence costs.	Developing cooperative solutions to problems can help to restructure elements of security relationships, reflecting altered ways of understanding each other and the nature of security relations.
CONFIDENCE EXPANDING	Develop cooperative practices and principles for the coordinated, more effective management of security threats.	Generally cooperative (primarily non-zero-sum thinking) but with some competitive bargaining (especially about resource allocation and measure application).	Development of cooperative practices and principles for coordinated and effective management of common security problems.	Developing coordinated solutions to common problems can restructure elements of security relationships and enhance collaborative capabilities.
UNCONVENTIONAL	Develop more cooperative practices and principles for the improved management of threatening relations.	Unclear but likely competitive (generally zero-sum).	Generally improved relations with neighbours. Resolution of specific irritants or threats.	Unclear. Could resemble any of above.
RHETORICAL	Participate in order to: Appear to be a serious international actor; gain other objectives as trade-off; or to weaken an adversary.	Unclear but probably not relevant given absence of commitment to constructive outcome.	Image-enhancement; attainment of other (non-related) objectives; or the relative weakening of an adversary .	Discussing confidence building will look like the "real thing." Skillful diplomacy can deceive adversaries or attain other goals.

APPENDIX III – CONTINUED

CONFIDENCE BUILDING VARIANT	PERCEIVED NATURE OF THREAT ADDRESSED BY CONFIDENCE BUILDING	SUPPORTING CONDITIONS	UNDERLYING THEORY OF INTERNATIONAL RELATIONS	VARIANTS
IMPOSED	Use of violence and coercion by target state.	Target state must be subject to virtually total control and not in a position to resist application of measures.	Realist (balance of power).	A: Focus on straightforward imposed transparency and constraint; B: Focus on above but with hope/expectation of target state rehabilitation.
COMPETITIVE CONSTRAINT	Surprise attack; inadvertent conflict due to misperception.	Minimal. Sufficient interest and incentive to negotiate. Perhaps prolonged absence of conflict.	Realist (balance of power).	A: Transparency; B: Transparency and constraint.
BASIC	Inadvertent conflict; unnecessary tension and defence costs; surprise attack.	Some shared sense of concern and emergent dissatisfaction with status quo security policies and their costs. Perhaps new leadership. Possibly a key triggering initiative.	Realist but perhaps with some emerging sense of limited institutionalist possibilities.	A: Transparency; B: Transparency and constraint; C: Transparency with non-traditional security focus; D: Transparency and constraint with non-traditional security focus
ENHANCED	Unnecessary tension and defence costs; inadvertent conflict.	Growing sense that at least some security concerns require cooperative solution. Serious dissatisfaction with status quo approaches and costs. New leadership and perhaps a key triggering initiative.	Probably a combination of realist and institutionalist ideas (effectively, liberal institutionalist) with a growing sense of institutionalist possibilities.	A: Traditional security focus; B: Non-traditional security focus; C: Advanced form with traditional security focus; D: Advanced form with non-traditional security focus.
CONFIDENCE EXPANDING	Security problems that are difficult to manage unilaterally.	Good existing relations. Common problems best addressed in co-ordinated fashion. Willingness to adopt collaborative approaches and (perhaps) spend more.	Probably a combination of realist and strong institutionalist ideas (liberal institutionalist) with a strong sense of institutionalist possibilities.	A: Traditional security focus; B: Non-traditional security focus.
UNCONVENTIONAL	Unconventional threats that contribute to tension and suspicion or threaten well-being of state.	Unclear.	More likely realist but institutionalist sensibilities also possible.	Uncertain but some variety is possible.
RHETORICAL	Not really applicable. Participation entails an opportunity to take advantage.	A willingness (on the part of others) to engage in the process.	Realist.	A: Image-conscious; B: Quid pro quo; C: Deceptive

APPENDIX IV — TRADITIONAL, NON-TRADITIONAL, AND UNCONVENTIONAL SECURITY CONCERNS[11]

Confidence building normally is associated with traditional military concerns (problems associated with the capabilities and activities of traditional armies, navies, and air forces), but there are two other types of security concern that may also benefit from confidence building efforts. Practical usage suggests that this is the case, particularly with respect to non-traditional security threats. It is important to understand these extra dimensions because they can expand the scope of activities where confidence building can play a constructive role.[12] This may help to isolate the key or defining aspects of authentic confidence building by broadening the analytic focus. It also may help to identify appropriate limits for the confidence building approach.

Non-traditional security refers to a collection of problematic issues that extend beyond traditional security concerns but whose management still entails the participation of security personnel, broadly conceived. These are state actors authorized to use force and with responsibilities that include an extra-national component.[13] The security issues that can or do produce suspicion, misperception, or concern about the use of force are problematic in themselves (such as drug trafficking) or become problematic when other states use security forces to deal with them in troubling ways (such as the intrusive, "hot pursuit" of drug traffickers). This expands the scope of institutional participation to include, for instance, coast guards, immigration and customs, fisheries patrol, other resource management and pollution control agencies, port authorities, intelligence agencies, and police. Important examples of non-traditional security issues include arms trafficking, drug trafficking, piracy, international terrorism, undocumented migration, and environmental and resource management concerns.[14]

Unconventional security issues lie beyond even this extended boundary but still entail threats to state security in the broadest sense. As with non-traditional security concerns, at least some important aspect of these types of threat has an extra-national character in terms of origin or impact and can adversely affect the fundamental well-being or sovereignty of a state. However, unconventional security threats entail

---

11. This material is drawn from James Macintosh, "Diverse Interpretations of Confidence Building," a paper prepared for Canada's Department of Foreign Affairs and International Trade (November 1997). It also appears in *Non-Proliferation, Arms Control and Disarmament: Enhancing Existing Regimes and Exploring New Dimensions*, ed. By Peter Gizewski (Toronto: Centre for International and Security Studies, York University, 1998).
12. This will become increasingly important if states expand the scope of responsibilities normally associated with their militaries. This is likely to happen if traditional security concerns recede in importance.
13. This means that the state actors in question have responsibilities that include managing or monitoring the activities of foreign nationals, entities, and/or government agents that represent a threat to national security, typically within a state's own "space."
14. Note, however, that not every aspect of these agencies' concerns counts as a non-traditional security concern. Controlling the smuggling of parrots may be a serious issue but it does not count as a security issue. Controlling the smuggling of drugs does. Regulating the activities of domestic fishing vessels does not count as a non-traditional security concern but regulating the activities of foreign fishing vessels in domestic or controlled waters does.

no direct component of action that involves state actors authorized to use force. If they do entail the use of force (for instance, the extra-national persecution of ethnic or religious minorities that have populations extending beyond one state's border — perhaps because these communities are thought to harbour terrorists), then they fall into the category of non-traditional security threats.

One example of an unconventional security threat is significant, harmful trans-border pollution-producing activity by a neighbouring state. This is activity (typically economic) that entails the indirect deposition of pollutants, primarily as a side effect, within another state's territory. The upstream contamination of rivers with agricultural or industrial residues illustrates this type of threat, as does the use of coal-fired power plants that deposit acid rain and other pollutants downwind. Another example of an unconventional security threat is the use by one state of shared or common resources, such as water, to the point that other states can no longer meet their own basic needs. The economic policies or activities of foreign actors (states, cartels, or corporations) that fundamentally threaten the well-being of vulnerable states represents yet another example of an unconventional security threat. The threat to the banana industries of some Caribbean states (the overwhelming source of their export revenues) posed by international trade negotiations and the policies of fruit-exporting corporations provides a good illustration of this type of threat.[15]

For the purposes of this chapter, the distinction between non-traditional security threats and unconventional security threats is a more useful construction than a concept like "comprehensive security" which combines both sets of issues.[16] Most important, this distinction makes it easier to include new types of security issues (for instance, non-traditional security issues) that seem amenable to recognizable confidence building solution without automatically including the more problematic class of unconventional security concerns.

---

15. The notion of an "unconventional" security threat is not entirely satisfactory for several reasons, including the potential elasticity of its boundaries. For instance, it could be argued that the content and tone of high school and university history textbooks in neighbouring states represent a real, if unconventional, security threat if they foster negative images or breed general suspicion about national motives and intentions. Should this also be considered a type of security threat? Where do we draw the line? This general type of security threat is included here because at least some states interested in "confidence building" have identified it as a valid type of security issue treatable with confidence building-type solutions. This terminology is preferable to "non-military security threat" which comes close to meaning "non-security security threat" and blurs the nature of non-traditional threats.
16. "Comprehensive security" is a broad concept that embraces "economic, social, political and military cooperation; the development of mutual trust through military confidence building and the lowest possible level of armaments; the peaceful settlement of differences; open markets; transnational issues cooperation, such as transportation, communications, energy, science and technology, environmental protection, human migration, combating crime and terrorism; and a preparedness to contribute to security in neighbouring areas." (From an untitled, unofficial Foreign Affairs working paper. This view does not necessarily reflect official Canadian Government policy.) For some, this very broad concept is synonymous with the term "non-traditional security," although this is not the understanding employed in this paper.

Chapter Eight

A Total Ballistic Missile Ban Treaty

Phillip J. Baines*

Introduction

An advertising guru once quipped, "Nothing sells better than death and sex." At the end of the Cold War we find ourselves with an industrial capacity to produce aerospace weapons and a military hierarchy far in excess of what is necessary to defend our respective interests. Like the plot line of a good conspiracy novel, we can see a global military complex trying to manufacture the market "pull," and a global industrial complex trying to develop the technology "push," that would fill this idle industrial capacity and provide job security for industrialists, generals, and politicians. The result is a collection of interests that would have us buy "death" in the form of new and improved ballistic missiles, and "sex" in the allure of exotic technology that can defend against these new and improved ballistic missiles. Just like that other product, cigarettes, however, we must be wary of our addiction to an arms race and recognize the fact that we do not need to buy the agents of our destruction nor do we need to buy the mirage of the masculine cowboy in the post-Cold War world.

Accelerating calls for nuclear disarmament and the natural obsolescence of deployed strategic systems will present us with an opportunity to "kick" our addiction to nuclear weapons and their delivery platforms. As nuclear disarmament within the START process progresses to ever lower numbers of nuclear weapons, a total ballistic missile ban treaty becomes an increasingly attractive measure to increase international security and decrease expenditures for national defence. Simultaneous with these developments is the natural aging of deployed ballistic missiles that will soon reach the end of their operational lives, beyond which confidence in their ability to perform as specified will increasingly come into question. Like never buying the next package of cigarettes, it is possible to escape from this life-threatening addic-

---

* This chapter is an edited version of a paper commissioned for the Fifteenth Annual Ottawa NACD Verification Symposium, "Arms Control and the Rule of Law: A Framework for Peace and Security in Outer Space," 11–14 March 1998, Montebello, Quebec, Canada. The opinions expressed in this chapter are those of the author and do not necessarily reflect those of the Department of Foreign Affairs and International Trade or the Government of Canada.

tion though the exercise of sheer will power by refusing to buy replacement missiles. A total ballistic missile ban will not be "cold turkey" nuclear disarmament, as airborne delivery systems will still remain, but it will assist with the timely execution of obligations to comply with Article VI provisions of the Non-Proliferation Treaty for nuclear disarmament.

In the absence of such an arms control agreement, it is likely that billions of dollars will be spent in a renewed arms race spurred by the need to invest in both additional offensive ballistic missiles (to defeat new defensive systems) and in ballistic missile defensive systems (to defeat improved offensive systems). However, by utilizing the resources of improved space surveillance sensors tracking all aerospace objects in flight above the Earth and numerous Earth remote-sensing satellites viewing all terrestrial activity below, together with the demonstrated effectiveness of on-site inspections, it will become possible to verify a complete ban on the development, testing, deployment, and use of ballistic missiles and ballistic missile defence systems. What is required by the international community is the political will to suffocate the technology that fuels the nuclear arms race.

The Arms Race Phenomenon

One way to illustrate the arms race is to plot offensive and defensive weapons developments in a matrix to illustrate the coupled relationships amongst these weapons as follows:

	Strategic Arms Development	
	Offensive Weapons	**Defensive Weapons**
Offensive Weapons	**X**(2),(6)	**X**(1),(5)
Defensive Weapons	**X**(3)	**X**(4)

The arms race then proceeds something like this: (1) the presence of superior air, sea, and land defences, and the absence of effective ballistic missile defences; (2) spurs weapon designers to develop ballistic missiles; (3) the development of ballistic missiles; (4) spurs development of ballistic missile defences; (5) the appearance of ballistic missile defences; (6) spurs production of more ballistic missiles; and so on. This is an illustration of a quantitative arms race.

The arms race can also be qualitative, wherein the presence of ballistic missile defences (5), leads to the development of multiple independent reentry vehicles (MIRVs) (6), to saturate the defences (4), and/or, the development of manoeuvrable reentry vehicles (MARVs) (6), to penetrate the defences (4). The qualitative arms race arises from the improved quality of the boosters to carry more than one warhead and the improved quality of the reentry vehicles to evade or penetrate the limited defences.

In design theory, such a relationship amongst the variables is termed a "coupled" system. It is well known that progress in product development can only occur through two options in such situations. The first is to decouple the relationships through choices in design that essentially removes or diminishes the strength of an off-diagonal entry in the matrix. The second option is to treat the whole problem in an entirely new way for which solutions are well known. Any other course of action leads to a stagnation in the design process with the attendant risk that later in the product cycle the design solution will fail to function as it should.

The ABM Treaty was an attempt to use the first design method to solve the coupled problem of ballistic missiles and ballistic missile defences. The negotiation of the ABM Treaty effectively reduced the strength of the coupling of the non-diagonal elements of the strategic arms matrix by placing restrictions on the effectiveness of the defensive system (upper right-hand entry of the matrix). The simultaneous negotiation of the Strategic Arms Limitation Talks (SALT I) Agreement effectively reduced the strength of coupling of the non-diagonal elements of the strategic arms matrix by limiting the number of delivery systems any defensive system would encounter (lower left hand entry of the matrix). It was thought that this would lead to a stable environment in which nuclear disarmament could proceed, or, at least, place some manageable cap on the number of strategic weapons required for deterrence on both sides of the Cold War. A testament to the strength of coupling in non-diagonal elements of the strategic arms development matrix and the weakness of the SALT I Agreement, which only limited delivery systems as opposed to warheads, spurred development of MIRVs. The consequence of MIRVs was increased ballistic missile performance to carry more warheads and penetration aids, and, to propel research into the Strategic Defense Initiative (SDI) that would effectively abrogate the ABM Treaty prohibiting space-based ABM systems to better defend against these improved ballistic missiles.

At the end of the Cold War we have the opportunity to try the second design method, also known as the "Alexander the Great Method." When Alexander the Great encountered the puzzle of the Gordian knot, he was asked if he knew the solution to the puzzle. Withdrawing his sword, he quickly cut the knot. Alexander aptly demonstrated his knowledge of the second design solution to complex problems as being one of redefining the problem with new parameters that one already knows how to solve. Thus, both history and theory dictate that ballistic missiles and ballistic missile defence systems will have to vanish simultaneously. This design solution, negotiation of a total ballistic missile ban treaty (TBMBT), will see the simultaneous elimination of ballistic missiles and ballistic missile defence interceptors. Proper emphasis on the problem posed by ballistic missiles and ballistic missile defence is, therefore, to "remove" the threat and not to "defend against" the threat of ballistic missiles. With no ballistic missile threat there is no need to create ballistic missile defences. This solution in effect collapses the strategic arms matrix for this class of weapons. Ballistic missile defence interceptors must also be eliminated because the commonality of design solutions and mission requirements give anti-ballistic missile interceptors the capability to carry either interceptors or single reentry vehicle warheads. While technology has yet to give a viable solution to the ballistic missile threat, the international community has a proven ability to negotiate comprehensive arms control agreements. The challenge before the international community is to perform this feat before the expanse of ballistic missile and ballistic missile defence deployments leads to a new arms race, or worse, a preemptive strike against such deployments.

Constituents of a Total Ballistic Missile Ban Treaty

One proposed framework of a TBMBT might include the following provisions:

- open to signature by all United Nations members;
- all parties undertake not to develop, test, deploy, or use ballistic missiles or fractional orbital bombardment systems with ranges in excess of 150 kilometres;
- all parties undertake to dismantle all ballistic missiles and fractional orbital bombardment systems within five years of entry-into-force;
- all parties undertake not to develop, test, deploy, or use ballistic missile and satellite interceptors with capabilities in excess of a 40-kilometre interception altitude and a 2 km/s interceptor velocity;
- all parties undertake to dismantle all ballistic missile and satellite interceptors within five years of entry-into-force;
- all parties undertake not to transfer missile or interceptor technology, components, or systems to non-parties;
- all parties may develop, test, deploy, and use space-launch vehicles and sounding rockets, provided they do so under a civilian authority and are subject to specified confidence building measures;
- all parties shall have non-discriminatory access to space-launch services of the parties providing the launch of such satellites does not endanger international peace and security;
- entry-into-force occurs when all nations with space-launch vehicles and ballistic missiles have ratified the agreement;
- a complement of declarations, on-site inspections, terrestrial-based outer space surveillance systems, space-based in-orbit monitoring, and space-based Earth imaging systems shall verify a suitably framed agreement;
- economic penalties in the space sector will apply to those nations choosing to remain outside of the treaty. Economic sanctions will apply to those state parties found in material breach of the treaty. Security Council-authorized actions may apply subsequent to this determination;

- spacecraft science and technology transfers not associated with space-launch vehicles may be freely exchanged by state parties to develop satellite payloads for scientific research and commerce;
- necessary and sufficient data selected from national technical means (NTM) of verification is permitted as admissible evidence for compliance determinations;
- an international organization is required to administer this treaty;
- ultimate compliance determinations, however, reside with the state parties; and,
- the treaty is of unlimited duration with reviews every five years and without the ability to withdraw.

Rationale

The rationale for the TBMBT arises from the realpolitik that continued nuclear disarmament with lower overall levels of nuclear weapons is not possible in the presence of ballistic missile defences (BMD), since a credible nuclear deterrent may require a significant number and high quality of ballistic missile reentry vehicles to saturate any defensive system.

Any introduction of BMD systems will invariably:

- derail reductions for those countries with significant numbers of ballistic missiles; and,
- buildup the arsenals of those countries with limited numbers of ballistic missiles.

For nations to do otherwise is to accept certain defeat with the attendant consequence of being coerced by nuclear threats from a single nation forearmed with both a sword and a shield. The former case has been advanced by Russia while the latter has been put forth by China. In the worst case, preemptive actions or asymmetric responses that also require vast defence expenditures will be the likely outcome of any decision to deploy substantial ballistic missile defence systems. The consequence of the introduction of BMD systems will then delay progress in nuclear disarmament by the five declared nuclear weapons states. Any delay in nuclear arms reductions will at least severely strain negotiation of further non-proliferation agreements and may, at worst, cause the breakup of the existing Non-Proliferation Treaty regime. This is the South Asian and Middle East dimension of the problem.

The Missile Technology Control Regime (MTCR) may not be able to stop ballistic missile proliferation indefinitely and may only slow down the inevitable acquisition of an indigenous capability by several nations. A treaty that could encompass all nations, including Iran, North Korea, India, and Pakistan, would correct the inadequacies of non-universal supplier control regimes to stem the flow of ballistic missile technology.

As the number of nuclear weapons falls below 2,500 nuclear warheads (START III levels), the assumption of a nuclear triad comes under increasing scrutiny. Indeed, some believe that post-Cold War security requirements could be served by a mere 1,000 nuclear warheads for the United States and Russia if the remaining three declared nuclear weapons states maintain their existing force levels. At this level of armaments, real choices must be made among the services to determine which will carry the nuclear deterrent with its attendant power in terms of national defence appropriations.

During the Cold War, the most threatening systems were judged to be land-based ballistic missiles with MIRVs. These systems were the most dangerous because they had the highest accuracy, could carry the most reentry vehicles, and had the greatest throw-weight. This latter feature enables ballistic missiles to fly additional penetration aids to confuse defensive systems. These qualities made intercontinental ballistic missiles (ICBMs) the ideal first-strike weapon. When reusable single-stage-to-orbit launch vehicles come into operational use, ballistic missiles will no longer have a dual-use capability. The economic competition from these new space-launch vehicles will make ballistic missile technology too expensive to capture any appreciable market share of the commercial space-launch industry. Consequently, land-based ballistic missiles may become a prime candidate for the scrap heap if the end purpose is to enhance international security.

Sea-launched ballistic missiles (SLBMs) pose a different threat to international security. The closest the world came to annihilation since the Cuban Missile Crises was a launch of a Norwegian sounding rocket mistaken for an American SLBM. SLBMs have the least warning time and their accuracy has improved to the point of making them an ideal first-strike weapon with which to decapitate the command and control infrastructure of an enemy. Unlike ICBMs, SLBMs have never enjoyed a dual-use purpose and so they too become an ideal candidate for elimination from future force structures. Counteracting this end result is the fact that submarines are the most secure leg of the nuclear triad in that their stealth ensures that they can ride out a first-strike and be available for a retaliatory strike as called for under nuclear deterrence doctrines.

Bomber aircraft are the slowest means for delivering nuclear weapons and are recallable. As such, air-based nuclear weapons are the most stable form of nuclear deterrent. In addition, air defence systems are the strongest element of defence in any nation's arsenal. Consequently, the expected loss of air crews and the uncertainty in mission success might raise an additional barrier to nuclear adventurism if strategic and tactical aircraft were the only means to deliver a nuclear weapon.

Elimination of ballistic missiles, therefore, would maintain a balance of nuclear forces between all American services. That is to say, the US Air Force would lose ballistic missiles but would retain bombers, the US Navy would lose ballistic missile submarines but retain aircraft carriers, and the US Army would lose ballistic missile defence but would retain an enhanced mandate for improved air defences. The expected savings from military expenditures could amount to billions of dollars for the US alone. Comparable savings for China and Russia would be likely.

History also illustrates that as defence budgets wane triad structures are rationalized. In France, for example, the triad structure of "force du frappe" was reduced to include only ballistic missile submarines and nuclear-capable aircraft. The United Kingdom never developed a triad force structure and has recently eliminated the Royal Air Force from the British deterrent by retiring the nuclear gravity bombs from its arsenal. The Royal Navy is now the only service that carries Britain's nuclear deterrent. Both Russia and China maintain triad structures but would likely want to reduce military expenditures further to fuel economic growth in order to absorb redundant military personnel. China also has an interest in improving relations with the United States in order to secure access to technology. Table 1 illustrates the current force structures of Russia and the United States. Should ballistic missiles be dropped from their inventories, a force structure appropriate to today's threat environment would be retained.

Table 1. American and Russian Strategic Force Structures

Platform	American Total Warheads[a]	Russian Total Warheads[b]
ICBMs	2,000	3,610
SLBMs	3,456	1,824
Aircraft	1,800	806
Total	7,256	6,240

a R.S. Norris and W.M. Arkin, "U.S. Strategic Nuclear Forces, End of 1997," *The Bulletin of Atomic Scientists* (January/February 1998), 70–72.

b R.S. Norris and W.M. Arkin, "Russian Strategic Nuclear Forces, End of 1997," *The Bulletin of Atomic Scientists* (March/April 1998), 70–71.

Finally, a TBMBT would be the natural conclusion of the detargeting, dealerting, and demating deliberations currently under study. An air-based deterrent possesses all of the desirable features of these options. An elimination of ballistic missiles and ballistic missile defences would increase international security and decrease national expenditures on defence.

Verification

Former US President Ronald Reagan had said that any arms control agreement must possess three qualities. It must first be equitable. Elimination of all ballistic missiles and ballistic missile defence systems in the world would meet this requirement. The second requirement is that it must be verifiable. The verification measures required for the TBMBT would not be dissimilar to the provisions in the START, INF, and ABM treaties. Measures exploiting national technical means of verification, on-site inspections, and ongoing monitoring and verification measures at production facilities are all proven methods that can be applied to the verification of an TBMBT.

In addition to past verification technologies, imminent technologies will become available to verify a TBMBT. Before a missile interceptor may defeat an incoming reentry vehicle, the target must be detected and tracked with sufficient accuracy to enable guidance for a "hit-to-kill" intercept. The next generation space-based early warning and space surveillance sensors, such as the US SBIR-High and SBIR-Low systems, will provide the capability to track all aerospace objects ascending through the atmosphere and operating in the lower altitudes of outer space. The existing Mid-Course Space Experiment (MSX) satellite is demonstrating this space surveillance ability just as upgrades to the existing Defense Support Program (DSP) satellites are providing improved warning of theatre-range ballistic missile launches.

Following these technological advances is the trend to share what was once highly classified data with an expanding community of nations. For example, early-warning information is already exchanged between the United States and Canada, NATO, Japan, and Israel. NORAD SpaceTrack catalogues have been exchanged with Russia. Additionally, former US presidents had offered to share even Strategic Defense Initiative (SDI) and Global Protection Against Limited Strikes (GPALS) technology with the Soviets. As concern over accidental or unauthorized release of Russian strategic nuclear forces mounts in the West, the US may need to share early-warning data with Russia in real-time.

The TBMBT will also need to address a ban on fractional orbital bombardment systems that will look and behave strikingly similar to commercial micro-gravity return capsules. Here improvements in the capabilities of terrestrial-based tracking and identification of satellites, like those which the US Advanced Electro-Optic System (AEOS) will soon afford, may also help in the verification of a TBMBT. Clearly then, those capabilities needed to perform the ballistic missile defence mission give rise to the ability to verify a TBMBT.

Finally, the last of the above-mentioned requirements of an arms control agreement is that it must be in a nation's national security interest. Eliminating a global threat posed by ballistic missiles and ballistic missile defences, is in the interest of all members of the community of nations.

Significance for Outer Space

A TBMBT would eliminate the threat of conventionally-armed missile interceptors reaching into outer space to threaten satellites in orbit about the Earth. A TBMBT would also enhance prohibitions of the Limited or Partial Test Ban Treaty (LTBT) and the Comprehensive Test Ban Treaty (CTBT) against test and use of nuclear-armed missile interceptors capable of reaching into outer space. As such, a TBMBT would complement a possible convention for the non-weaponization of outer space[1] that prohibits space-based weapons threatening targets in orbit or on the Earth.

Exploiting the verification abilities of a TBMBT would facilitate the creation of a space traffic control architecture and a secure international environment that could enable transoceanic ballistic missile flight for civilian passenger and cargo traffic on reusable single-stage-to-orbit launch vehicles. Failure to pursue this course of action will see the foreclosure of the pursuit of a civilian outer space regime in favour of military spaceplanes and ballistic missile defence systems. Clearly, economic interests would be better served if single-stage-to-orbit vehicles become as widespread as today's jumbo jetliners, instead of a future that holds only one or two squadrons of military spaceplanes and a couple of hundred ballistic missile defence systems.

Conclusions

The introduction of BMD systems poses an unacceptable risk to successfully concluding a nuclear disarmament agreement. In order to fulfil the Article VI provisions of the NPT, the elimination of ballistic missiles simultaneously with ballistic missile defences will be required. A total ballistic missile ban treaty (TBMBT) may accomplish this Herculean task, given sufficient political will. The consequence of not acting on nuclear disarmament is the risk of a complete rupture of the non-proliferation regime. Given that technology has yet to provide a defence against ballistic missiles, this rupture would be a dangerous one.

Without an elimination of ballistic missile defence systems, the international community cannot obtain a meaningful anti-satellite prohibition to improve peace and security interests in outer space. Prior to negotiation of a TBMBT, however, the antecedent convention for the non-weaponization of outer space could serve as the catalyst for achieving a TBMBT by establishing the necessary momentum in this direction. The consequence of these twinned agreements would be the preservation of outer space for peaceful purposes and undiminished security at reduced levels of armaments.

---

1. Phillip J. Baines, "A Variant of a Mandate for an Ad Hoc Committee on Outer Space within the Conference on Disarmament: A Convention for the Non-Weaponization of Outer Space," see chapter six of this volume.

Chapter Nine

The Protection of Satellites: Problems for Arms Control

George Lindsey*

Existing Treaties Involving Weapons in Space

The multilateral Outer Space Treaty, signed in 1967, prohibited the placing in orbit around the Earth of any objects carrying nuclear weapons or any other weapons of mass destruction. This did not ban ballistic missiles or "Fractional Orbital Bombardment Systems," which transit through space for a few minutes and then return to Earth, before the completion of one orbit. The treaty contained no provision for verification.

The bilateral ABM Treaty, signed in 1972, limited the numbers and performance characteristics of defensive AICBM (anti-intercontinental ballistic missiles) systems. This included the prohibition of the development, testing, or deployment of sea-based, air-based, or space-based AICBM systems and their components. It was agreed that in the event of future ABM systems employing "other physical principles," limitations on their use would be discussed between the Parties.

The only concession to a need for verification was an agreement not to interfere with national technical means (NTM) or use deliberate measures of concealment. This provision has significant implications for space, since the primary form of NTM used for verification has been, and still is, space surveillance.

Defence Against Theatre Ballistic Missiles (ATBM or TMD)

Interception in the Mid-Course and Terminal Phases

The problem of intercepting a theatre ballistic missile (TBM) in its mid-course or terminal phase has very different parameters from the AICBM problem. In Figure 1, the top diagram shows the full trajectory of a 10,000 kilometre ICBM; the lower diagram depicts (on a larger scale), the last 2,000 kilometres of the ICBM trajectory, and the complete trajectory of a 500 kilometre short-

---

* This chapter is an edited version of a paper commissioned for the Fifteenth Annual Ottawa NACD Verification Symposium, "Arms Control and the Rule of Law: A Framework for Peace and Security in Outer Space," 11–14 March 1998, Montebello, Quebec, Canada.

range ballistic missile (SRBM). The hatch marks on the trajectories indicate the number of minutes before the missile will impact. The line of sight from a sensor located on the ground near the impact point is shown by the dashed line marked L.O.S.

It can be seen that an AICBM system located on the ground near the impact point will not have a chance to detect and begin to track the ICBM until it has completed about two-thirds of its trajectory, and is within eleven minutes of impact. In contrast, the SRBM will rise above the line of sight of a TMD sensor located on the ground near the point of impact very soon after liftoff — about five minutes prior to impact.

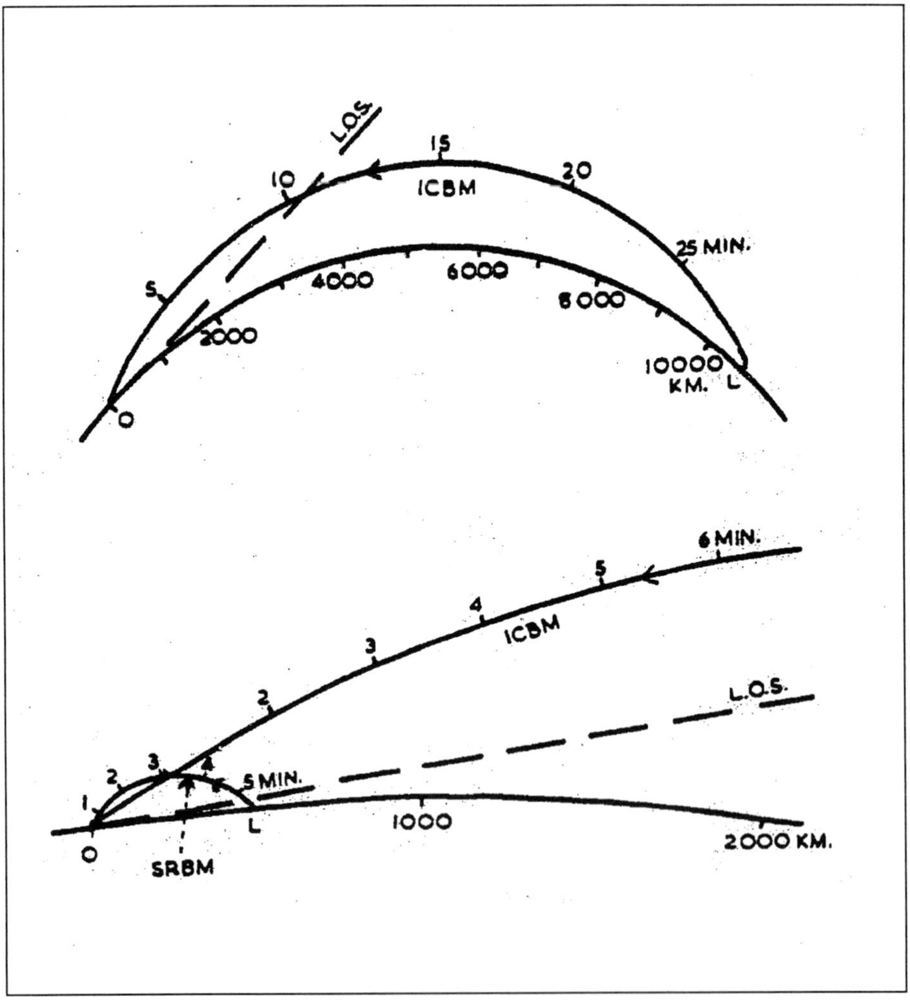

Figure 1

Table One lists some of the numbers associated with the trajectories, and adds those for SRBMs with maximum surface-to-surface ranges of 300 and 1,000 kilometres.

Table One: Surface-to-Surface Ballistic Missile Trajectories

	MAXIMUM RANGE			
Parameter	**300 km**	**500 km**	**1,000 km**	**10,000 km**
Target velocity (km/sec)	1.7	2.2	3.0	7.2
Trajectory apogee (km)	150	250	350	1,440
Total Time of Flight (minutes)	4.5	5.5	8.0	33
Time first visible above ABM battery horizon (minutes)	4.0	5.0	7.0	11

Sensors placed up-range of the AICBM battery (or, better still, at a high altitude) and able to detect the approach of ICBMs (or, better still, to track their precise positions) would provide surface-based AICBM defences with more warning time, allow each battery to protect a much larger "footprint" of the territory to be defended, and, perhaps, make it possible to "shoot-look-shoot" against each missile.

Spaceborne sensors could contribute vital assistance to surface-based AICBM defence. However, they would be somewhat less of an asset for surface-based defence against TBMs in the mid-course or terminal phase of their trajectories, during which tracking can be accomplished from the ground.

Early Warning and Interception in the Boost Phase

While a re-entry vehicle of a ballistic missile offers a small target for radar or visual sensors and a rapidly cooling one for infrared as it passes through empty space, the exhaust plumes from the booster rockets emit radiation easily detectable by IR sensors as far away as the 36,000 km altitude of a satellite in geosynchronous orbit (GEO). Early warning of a launch anywhere on the Earth can be provided by space-based sensors.

During its boost phase the propulsion stages of a ballistic missile are very vulnerable. The fuel tanks still contain unburnt rocket fuel and premature shutdown or loss of fuel would shorten the trajectory. For attack of a missile in its boost phase the most appropriate BMD[1] weapon may be a powerful laser

1. In this chapter the term BMD (ballistic missile defence) will comprise AICBM,

able to project a finely-focussed beam through empty space at the speed of light. To be in line-of-sight to the climbing missile, and to minimize the absorption and disturbance to the laser beam while passing through the lower atmosphere, the laser BMD weapon should be carried in a satellite or an aircraft flying at high altitude.

Thus, the most promising applications of satellites for defence against ballistic missiles are for provision of early warning of the launch of either ICBMs or TBMs, for cueing AICBM and ATBM defences, and for fire-control data (and possibly launching of weapons) for AICBM systems.

The Protection of Satellites

Even though satellites may be able to provide support for certain destabilizing functions — conduct of offensive ground operations, pinpointing of targets, guidance of missiles, location of ships for attack by submarines, etc. — these are outweighed by the many functions which they perform which are clearly stabilizing.[2]

Quite apart from considerations of military stability, the services that are provided by space technology for a host of civilian functions are already indispensable and are sure to go on expanding in the future. Many of the services, such as observation of the Earth, navigation, communication, and weather forecasting, are used for both military and civilian purposes, and the satellites which provide them would be both indispensable for normal commerce and legitimate military targets in a conflict.

Problems for Arms Control

A major objective for arms control is to improve worldwide strategic stability. One step toward this goal would be to make satellites safe from attack.

Protecting Satellites in GEO

Almost all of the functions provided by satellites in GEO are of a stabilizing nature. Examples include launch warning, detection of nuclear explosions, and rapid worldwide communications.

The most likely means of attacking satellites in GEO appear to be by other satellites armed with ASAT laser weapons. It might be possible to prohibit powerful spaceborne (and perhaps airborne) laser weapons. But less powerful lasers will be needed in satellites (and aircraft) for legitimate purposes such as remote sensing, communications, and scientific research. It

---

ATBM, and TMD.

2. The question of whether BMD is stabilizing or destabilizing is a complicated one. The presence of defences could motivate those planning aggression to desist from acquiring ballistic missiles, or, conversely, to build more of them in order to overwhelm the defences. AICBM defence interacts with strategic deterrence and much depends on which targets are being protected. But ATBM defence should be considered to be, on balance, stabilizing.

would not be acceptable to ban all lasers and it would be necessary to be able to verify that the lasers that were in a satellite (or aircraft) were not capable of being used as weapons.

Protecting Satellites in LEO

Most of the satellites whose functions could be considered to be destabilizing operate in LEO (as do many that are stabilizing). Examples include the identification and determination of the precise location of targets, support for terrestrial operations of war, and missile guidance. Satellites in LEO could be attacked by surface-based weapons using components already designed for use in ballistic missiles, ABM systems, and space launching, but with guidance and warheads designed for an ASAT role.

Table Two shows some of the velocities associated with surface-to-surface ballistic missiles and with satellites. It also lists the velocities necessary for a vertically-launched rocket to attain the altitudes of LEO and GEO.[3]

To maintain a circular LEO orbit, a satellite must move with a velocity of about eight kilometres per second. This tremendous velocity (about 18,000 miles per hour — twenty-five times the speed of sound in the lower atmosphere) makes satellites extremely vulnerable to an ASAT weapon which causes the huge kinetic energy of the satellite to destroy itself if it runs into even one small inert object. All the ASAT system has to do is put up a cloud of small objects in the satellite's predictable path, and let it run into one of them.

This is an easier task than interception of a ballistic missile, provided that the ASAT system can wait until the satellite's orbit comes close enough to the ASAT launch site, and that the interceptor missile can reach up as high as the altitude of the satellite. The ASAT battery is rather like waiting for a moving target to come into view in a shooting gallery.

As indicated in Table Two, a surface-based ASAT rocket would require a burnout velocity of 4.1 km/sec to deliver a payload to an altitude of 1,000 km, thus covering the most occupied layers of LEO satellites. But even a 2.5 km/sec burnout velocity will get beyond a 300 km altitude — well into the LEO band. In contrast, the 36,000 km altitude of GEO requires a burnout velocity of more than 10 km/sec (nearly enough to escape from the Earth

3. The burnout velocity to which a rocket is able to boost its payload depends on the mass of that payload. If the propulsion stage (or stages) of a surface-to-surface missile were used to boost an ASAT warhead which was lighter than the warhead of the SSBM, the burnout velocity would be greater. SSBMs deliver their destructive power by high-explosive warheads weighing hundreds of kilograms, while the ASAT warhead uses the kinetic energy of the satellite to destroy itself in a collision with a mass of a few grams. Even if they have to have sensors, small deflection jets, and control circuitry for homing guidance, and deploy a large number of small masses into the path of the target satellite, it seems likely that ASAT warheads can be considerably lighter than missile warheads.

Table Two: Velocities Required for Certain Purposes

SURFACE TO SURFACE BALLISTIC MISSILE	
Maximum Range	**Burnout Velocity**
SRBM 300 km	1.7 km/sec
SRBM 500 km	2.2 km/sec
SRBM 1,000 km	3.0 km/sec
SRBM 10,000 km	7.2 km/sec

VERTICAL ASAT ROCKET	
To Reach Altitude of	**Burnout Velocity**
LEO 300 km	2.4 km/sec
LEO 500 km	3.0 km/sec
LEO 1,000 km	4.1 km/sec
GEO 36,000 km	10.3 km/sec
ESCAPE	11.2 km/sec

SATELLITE IN CIRCULAR ORBIT	
At Altitude of	**Orbit Velocity**
LEO 200 km	8.2 km/sec
LEO 300 km	7.8 km/sec
LEO 500 km	7.6 km/sec
LEO 1,000 km	7.4 km/sec
GEO 36,000 km	3.1 km/sec

forever), which is far beyond the launch velocities necessary for ICBMs or AICBM interceptors.

Thus, it appears that satellites in LEO are vulnerable to attack by surface-based systems using warheads designed for ASAT, boosted up to LEO altitude by rocket propulsion of the type already deployed for surface-to-surface missiles, BMD interceptors, and space launches.

The ABM Treaty

The ABM Treaty was designed to restrict the development of AICBM defences, but without hindering the development of anti-aircraft or ASAT defences. Any agreement that would be effective in preventing threats to satellites in LEO from surface-based weapons would have to deal with many of the problems of the types that were negotiated in the ABM Treaty. It would also have to place severe restrictions on ATBM defence. It would have to require much more stringent provisions for verification than were agreed in the ABM Treaty. Finally, it would need to be a multilateral treaty.

Although there are continuing debates as to the thresholds of demarcation between AICBM and ATBM, anti-aircraft, and cruise missile defence, one threshold which has been given serious consideration for tests (which could be monitored from space) is a velocity of 3 km/sec for interceptor missiles and 5 km/sec for target test vehicles. An ATBM system tested with interceptor velocities slightly below 3 km/sec could have the potential to form the basis of an ASAT system that would be effective against satellites in the lower LEO altitudes.

An ATBM system launched against a test target vehicle moving at less than 5 km/sec would not represent a full-scale test for interception of a satellite moving at 8 km/sec in LEO. However, considering that the time of appearance and the trajectory of a satellite would be known long before the ASAT missile was launched, the slower speed of the test target (and the fact that it was not in Earth orbit) might not prevent a useful and valid assessment being made of the performance capability of the system.

An earlier threshold, intended to demarcate testing of AICBM from anti-aircraft systems — known as the "Foster Box" — would have specified the limits to come at a velocity of 2 km/sec for the interceptor vehicle and an altitude of 40 km for the target vehicle. These limits would exclude ASAT systems.

Other ASAT Threats

Satellites in LEO are just as vulnerable to spaceborne lasers as are those in GEO — and more vulnerable to airborne lasers. However, an agreement to prohibit spaceborne and airborne laser weapons able to destroy satellites would also prevent their effective employment for BMD, whether against intercontinental or theatre missiles, in the boost, mid-course, or terminal phases of their trajectory. Satellites in GEO could be deactivated or destroyed by astronauts transferring from personed satellites. Satellites in LEO could be deactivated by an electromagnetic pulse.

Conclusion

The prohibition of all types of weapons in space would afford considerable protection to satellites in GEO. So would the prohibition of laser weapons in space. But it would probably not be possible (or desirable) to prohibit laser instruments.

Prohibition of weapons in space would leave satellites in LEO vulnerable to surface-based and air-based BMD systems, to conversions of ballistic missile or space launch boosters to an ASAT role, and to personed space vehicles.

A treaty for the comprehensive prohibition of all ASAT systems would have to be multilateral, could not be compatible with the continuing capabilities of states possessing effective BMD systems, and would have to deal with the possibilities of the use of SSBM or space-launch boosters for an ASAT role.

Amendments to the ABM Treaty that would prevent all use of surface-based AICBM systems for ASAT operations could not preserve their potential effectiveness for AICBM.

A new treaty — or amendments to the ABM Treaty — that would prevent the use of surface-based ATBM systems for ASAT operations would seriously reduce the potential effectiveness of ATBM defence for its intended purpose.

Chapter Ten

The Ballistic Missile Defence and Anti-Satellite Relationship

Phillip J. Baines*

Introduction

Arms control initiatives at the Conference on Disarmament and elsewhere have tried on many occasions to prohibit weapons based on functional definitions according to their respective missions. Initiatives such as anti-satellite bans in the presence of anti-ballistic missile systems, attempting to authorize space-based theatre ballistic missile defence systems while simultaneously prohibiting space-based anti-ballistic missile defence systems, or verifying bans on fractional orbit bombardment systems in the growing presence of commercial micro-gravity reentry capsules is problematic because of the shared technology and the shared requirements of the related missions. This chapter examines the relationship between the anti-satellite mission and the ballistic missile defence mission. The chapter concludes that a meaningful ban on the deployment of anti-satellite systems is not possible in the presence of deployment of either theatre or strategic ballistic missile defence systems. It is also suggested that efforts at the Conference on Disarmament regarding outer space must first focus on preventing its "weaponization."

Axiom

It is well recognized that a weapons system capable of defeating a ballistic missile reentry vehicle for ballistic missile defence (BMD) functionality is more than capable of defeating a satellite for an anti-satellite (ASAT) functionality because of the relative hardness of reentry vehicles versus satellites.[1] This conclusion arises from the differences in the targets and the

---

* This chapter is an edited version of a paper commissioned for the Fifteenth Annual Ottawa NACD Verification Symposium, "Arms Control and the Rule of Law: A Framework for Peace and Security in Outer Space," 11–14 March 1998, Montebello, Quebec, Canada. The opinions expressed in this chapter are those of the author and do not necessarily reflect those of the Department of Foreign Affairs and International Trade or the Government of Canada.

1. United States Congress, Office of Technology Assessment, *Anti-Satellite Weapons, Counter-measures, and Arms Control*, OTA-ISC-281 (Washington: US Government Printing Office, September 1985).

trajectories that each target follows in its respective flight regime. In the case of satellites, they may present larger target cross-sectional areas and are more fragile than reentry vehicles. Reentry vehicles may also move at higher velocities than orbiting satellites and may follow more erratic trajectories than the predictable orbits of satellites. This makes intercepting reentry vehicles more difficult than intercepting satellites.

The corollary is that an ASAT weapon would not necessarily make a BMD weapon because of the softness of satellites compared with the hardness of reentry vehicles. This might be particularly true in the case of directed energy weapons in which the aperture size and power output of lasers might enable national technical means of verification to distinguish those features needed for anti-satellite missions from those features required to penetrate the ablative material of reentry vehicles in a ballistic missile defence mission. A countervailing argument is that longer dwell times at lower powers can compensate for larger powers with shorter dwell times. Besides this feature of fundamental physics, the roughly equivalent hardness of satellites and the payload buses that carry the multiple reentry vehicles gives anti-satellite systems at least some residual capability for ballistic missile defence.

Yesterday

Throughout the development history of ballistic missile defence and anti-satellite systems, the two programmes have proved to be but opposite sides of the same coin. In the extreme case, for example, the nuclear-tipped Nike-Zeus anti-ballistic missile defence system was equally as capable of destroying satellites as it was at destroying incoming reentry vehicles. In fact, the propensity of nuclear-tipped interceptor missiles to take out "friend" and "foe" satellites alike, in addition to enemy reentry vehicles, has spurred the present research and development activities to pursue "hit-to-kill" strategies for ballistic missile defence. Even in this architecture, the BMD and ASAT design challenges and the design solutions are largely common for both missions. Each require boosters, guidance systems, sensors to discriminate the targets from the background, and both require largely common space surveillance systems and command and control facilities. It is not altogether unexpected then, that both BMD and ASAT systems would look strikingly similar to national technical means of verification.

In what has become the cornerstone of strategic stability between Russia and the United States, the Anti-Ballistic Missile (ABM) Treaty has restricted BMD for strategic-range ballistic missiles and has constrained air defence missiles from acquiring ABM capabilities. Prior to this watershed treaty, both defensive and offensive weapons systems were developed in lock step with one another. The ABM Treaty cut the Gordian knot of this arms race and replaced it instead with the concepts of deterrence and mutual assured destruction. This treaty enabled the two Parties of the day to place limitations on offensive ballistic missiles that otherwise would not have been possible.

Lacking any restrictions on defensive systems, a sufficient quantity of offensive systems must be retained to saturate any defensive systems in order to maintain a credible level of deterrence. But the logic of the ABM Treaty, to forgo defence in favour of limitations on offense, is counterintuitive to the military mind. Centuries of warfare taught that winning required both defensive and offensive capabilities in good measure. In a nuclear age, it became unthinkable that the only likely outcome of conflict would be a tie — a deadlock of strategic stability or a smoldering cinder of nuclear annihilation. The military mandate to ensure wins continues to oppose the arms control mandate to ensure ties.

Despite several attempts by US Presidents since Jimmy Carter, no similar agreement for constraints on ASAT development, testing, and deployment as those negotiated for anti-ballistic missile interceptors ever bore fruit. Before the recent demarcation agreement for the division of ballistic missile defence missions into theatre missile defence (TMD) systems and strategic ballistic missile systems, the lack of constraints on the ASAT mission could have provided an unconstrained avenue for ABM research and development — and possibly deployment — as long as the systems were never "tested in an ABM mode." Negotiations for constraints on ASAT systems could not be successful in this environment because of the deeply rooted psychological appeal of defending oneself from the threat of offensive systems.

Today

The recent demarcation agreement clarifying TMD activities has changed the military research and development environment in that development, testing and deployment of defensive systems, beyond those permitted for ABM systems, have become legitimate through agreement of the Parties to the ABM Treaty. With this development channel open it would now appear that an ASAT ban could be negotiated. Opposing this course of action, however, is the fact that TMD systems can also perform ASAT missions, as can National Missile Defence (NMD) systems under development by the United States. The NMD system of the United States is to be an ABM Treaty compliant system capable of deployment within three years of a decision to deploy that is made at least three years after an intense period of research and development begun in 1997. The system is not to be deployed unless the technology has proved capable of performing the mission and a recognized threat exists. Equivalent operational BMD systems also exist in Russia.

Figure 1 illustrates the capabilities of the American NMD and TMD systems in terms of altitude and closing velocity. Closing velocity is the speed and direction at which the target and interceptor approach one another. In the simplest case, two vehicles in the same orbit, but flying in direct opposition to one another, would have a closing velocity of twice the speed of any one vehicle. These capabilities need to be contrasted with the satellite population to illustrate the vulnerability of satellites to BMD systems.

Figure 1. Theatre Ballistic Missile and National Missile Defence Capabilities[2]

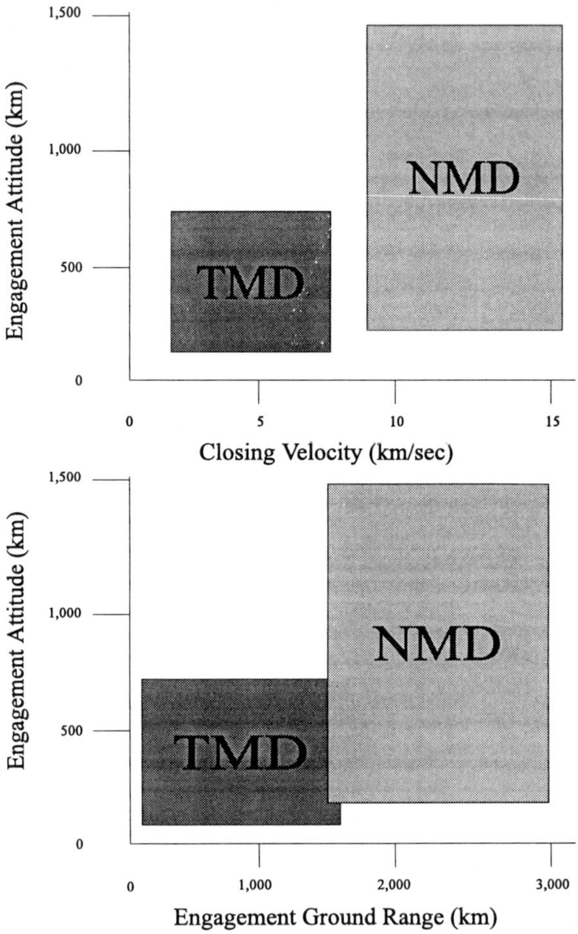

Figure 2 illustrates the satellite populations in low Earth orbit in terms of altitude above the surface of the Earth. The satellites most threatened by ASAT weapons are imaging and targeting satellites whose orbits occupy altitudes between 200 km and 1,500 km and whose velocities are between 7.7 km/s and 7.1 km/s respectively. Contrasting the capabilities of the TMD and NMD system capabilities in Figure 1 and the satellite populations in Figure 2 illustrates the capabilities of these BMD systems to perform ASAT missions. Further confirmation of these capabilities occurred with two recent events.

2. "U.S. Ballistic Missile Defence Program Focus," Department of Defence Fact Sheet.

The first occurred when the US President vetoed a dedicated kinetic energy anti-satellite (KE ASAT) system from the most recent defence appropriations bill without affecting that nation's national security.[3] The second occurred with the most recent launch of a Russian reconnaissance satellite into a higher orbit than the published capability of either of the US TMD or NMD systems.[4]

Figure 2. Satellite Population as a Function of Altitude[5]

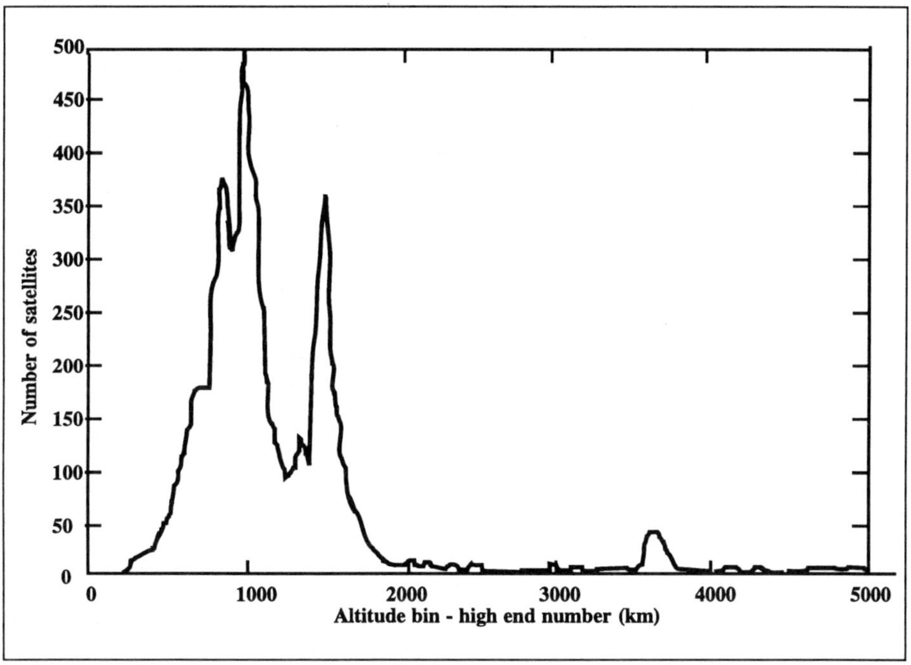

Because of this overlapping capability, an Earth-based, low altitude ASAT development, testing, and deployment ban is unattainable in the presence of a ballistic missile threat and TMD and NMD systems designed to defeat this threat. Similarly, an Earth-based high altitude ASAT development, testing, and deployment ban is unattainable in the presence of commercial launch vehicles and exo-atmospheric interceptor payloads developed for BMD. This is because it is but a small technical matter to swap the booster out

3. Press Briefing by OMB Director Frank Raines, NSC Senior Director for Defence Policy and Arms Control Bob Bell, and Deputy Defence Secretary John Hamre On the Line Item Veto, White House, Washington D.C., 14 October 1997 as available at http://www.whitehouse.gov.

4. C. Covault, "New Russian Recon Keyed to Area Surveillance," *Aviation Week and Space Technology*, 148:8 (23 February 1998), 37.

5. htttp://msx.nrl.navy.mil/msx_pi_teams/Pics/msx_space_surv.gif.

from under a successful exo-atmospheric kinetic kill vehicle to enable a more powerful space launch vehicle to attack satellites in higher orbits.

Concerning space-based weapons employed against satellites, existing agreements — such as the Limited or Partial Test Ban Treaty (LTBT), the Comprehensive Test Ban Treaty (CTBT), the Chemical Weapons Convention (CWC), the Bacteriological and Toxins Weapons Convention (BTWC) and the Geneva Protocol — may protect satellites from the "use" of weapons of mass destruction (WMD). The Outer Space Treaty may also protect satellites from the "use" of space-based variants of weapons of mass destruction through its prohibitions on deployment of such weapons in orbits about the Earth.

These multilateral restrictions may be reinforced by the recent demarcation agreements on BMD systems that prevent space-based weapons for TMD or ABM purposes among five State Parties. These restrictions would apply to both space-based missile interceptors and "exotic" weapons based on "other physical principles," such as space-based lasers that could substitute for missile interceptors. While these developments are positive, space-based ASAT systems are, however, not strictly prohibited. As such, ASAT development may still provide cover for space-based BMD research and development. One likely candidate is a "demonstrator" mission for a space-based laser planned for launch in 2005 by the United States.[6] Presumably using less power than a full-scale prototype weapon (prohibited by the ABM Treaty), advocates for the space-based laser "demonstrator" claim that such research is permitted. Any anti-satellite ban restricting development, testing, deployment, and use would further confound such research activities. Other critics of the demarcation agreement also point to how national governments, and not the Standing Consultative Commission, will now have the sole responsibility for deciding the compliance of systems with the ABM Treaty. This may prove to be a significant loophole for which ASAT, space-based laser "demonstrators," or "asteroid-busting" missions become suitable avenues to test space-based BMD architectures in contravention of the intent of the ABM Treaty.

Tomorrow

Given the evidence presented above, what might the future hold for improving the international peace and security environment for outer space? One subset of the problem might be addressed if space-based ASAT and BMD systems could be eliminated, with "space-strike" weapons (orbiting weapons that attack terrestrial targets), through a proposed convention for the "non-weaponization" of outer space that would at least ban the test and deployment of all space-based weapons regardless of purpose.[7] Table One illustrates the

6. J. Erlich, "BMDO, Air Force Boost Laser," *Defence News*, 12:40 (6–12 October 1997), 4.
7. Phillip J. Baines, "A Variant of a Mandate for an Ad Hoc Committee on Outer Space within the Conference on Disarmament: A Convention for the Non-Weaponization of Outer Space," see chapter six of this volume.

Table One. Multilateral Arms Control Matrix

Weapon's DEPLOYMENT Life-Cycle State	ENGAGEMENT SCENARIO (Weapon-Basing to Target-Basing)			
WEAPON CLASS	**Earth -to- Earth**	**Earth -to- Space**	**Space -to- Space**	**Space -to- Earth**
WMD	a,c	b,c	X[c,d]	X[c,d]
Conventional	e	e	d	d
"Exotic"	f	f	d	d

Key:

X (shaded)	Complete prohibition of weapon class with substantial multilateral participation.
(shaded)	Partial prohibition, or restriction(s) on weapon class with substantial multilateral participation.
(unshaded)	Unfettered freedom of action for a substantial population of the international community.

a Partial prohibition as CWC and BTWC only agreed to thus far. Nuclear element of WMD is still outstanding for weapons that "transits through" outer space.

b Partial prohibition as CWC and BTWC only agreed to thus far to prohibit deployment of WMD weapons that "reaches into" outer space. Nuclear component is still outstanding.

c The OST, LTBT, CTBT, CWC, BTWC collectively prohibit deployment of WMD that are space-based. Space-based means objects that complete at least one full orbit of any celestial body including the Earth, or on trajectories to or from any celestial body, excluding the Earth-to-Earth interaction, or stationed in outer space in any other manner.

d Potential entries as a result of a convention for the "non-weaponization" of outer space. Prohibition would also apply to the development, testing, and use life-cycle states of a weapon's product cycle.

e Potential entries as a result of a total ballistic missile ban treaty for weapons that "transits through" or "reaches into" outer space. Prohibition would also apply to the development, testing, and use life-cycle states of a weapon's product cycle.

f Potential agreement restricting Earth-based "exotic" weapons based on "other physical principles" like lasers that "reaches into" outer space. It might not be possible to differentiate between lasers capable of an anti-armour role and those capable of an anti-satellite role.

domain of the convention expressed in terms of weapon class and engagement scenarios between weapon-basing and target-basing options. Prohibition against development and use of space-based weapons besides test and deployment could be obtained with greater verification expense and intrusiveness.

Further ASAT arms control activities after successful negotiation of a convention for the "non-weaponization" of outer space might then require:

(1) negotiation of a total ballistic missile ban treaty to capture nuclear and conventional Earth-based weapons systems capable of "reaching into" outer space or "transiting through" outer space;[8] or,

(2) negotiation of a satellite immunity accord that essentially bans intentional harming of satellites through the "use" of Earth-based weapons systems.

The former option is more comprehensive in that it would prohibit development, testing, deployment and use of all ballistic missiles, fractional orbit bombardment systems and ballistic missile defence interceptors that could serve an ASAT role; whereas the latter case simply prevents the use of any weapons system against satellites. In either instance, a convention for the "non-weaponization" of outer space is a pivotal and an antecedent agreement as space-based weapons would need to be banned before any agreement could be reached to eliminate the means by which weapons in outer space could be removed by force.

Earth-based "exotic" weapons employing "other physical principles" — such as directed energy weapons, including lasers — would also require a suitable convention to complete the protection of satellites from all weapons classes that are Earth-based. As it might not be possible to differentiate lasers for anti-armour purposes from those for anti-satellite purposes, this agreement might need to be a comprehensive ban on all lasers in excess of a certain power level. This agreement would likely follow the two preceding agreements.

Conclusion

Several conclusions may be drawn from the relationship of anti-satellite and ballistic missile defence systems. The first is that an anti-satellite ban on development, testing, deployment, and use cannot come into existence until a ban is successfully negotiated for preventing the stationing of weapons in outer space. The rationale is that an anti-satellite-capable system is necessary to remove by force any attempt to control outer space with the deployment of space-based weapons.

The second conclusion is that an anti-satellite deployment ban may not be meaningful while there are ballistic missiles and ballistic missile defence

---

8. Phillip J. Baines, "A Total Ballistic Missile Ban Treaty," see chapter eight of this volume.

systems that can threaten artificial satellites. The threat posed by ballistic missile proliferation makes nations reluctant to forgo the promise of defence against these threats. The commonality between ballistic missiles, ballistic missile defence, and anti-satellite systems may require a sweeping agreement to remove all systems simultaneously in a total ballistic missile ban treaty.

The third conclusion sees the need for possibly three time-phased agreements to ensure international peace and security in outer space. The first agreement is a convention for the "non-weaponization" of outer space followed by a total ballistic missile ban treaty. A subsequent agreement for Earth-based "exotic" weapons, such as lasers reaching into outer space, would then complete the peace and security environment for outer space.

Chapter Eleven

The Role of Non-Governmental Organizations in Maintaining Space as a Weapons-Free Environment

Peter Stibrany*

This chapter makes some propositions regarding the potential benefits of two specific types of NGO involvement in a multilateral space arms control process. The two questions considered in particular are:

- Do NGOs have a role in creating visibility and momentum which could initiate a space arms control process and sustain it to a successful conclusion?
- Do NGOs have a role in monitoring a multilateral space arms control agreement, should one be concluded?

One should recognize at the outset that, by definition, the world of non-governmental organizations is extremely broad and covers interested parties with a wide variety of interests, agendas, and capabilities. Just categorizing these organizations and their characteristics would be a challenge. One distinction which is useful for the purposes of this chapter is that between NGOs which are principally advocacy groups and have little or no technical capabilities and NGOs which are principally technical in nature or have technically capable and active members. In answering the first of the questions posed above, no distinction will be made between these groups. The question of how to create visibility for space arms control and momentum to put this issue on the agendas of key governments is most heavily influenced by more general factors. In answering the second question, however, the significance of different NGOs is more important. The NGOs chiefly addressed in this section are the technical ones which observe the skies directly and catalog the activities of satellites. It also includes groups of technically competent people who, either professionally or as a hobby, comb through the published literature and keep track of what is being said about the space plans of governments and private enterprises around the world. It

---

* This chapter is an edited version of a paper commissioned for the Fifteenth Annual Ottawa NACD Verification Symposium, "Arms Control and the Rule of Law: A Framework for Peace and Security in Outer Space," 11–14 March 1998, Montebello, Quebec, Canada.

should be noted that some of these NGOs are not formal organizations at all; they can be as diffuse and unstructured as a set of individuals who monitor and contribute to USENET news groups.

One could reasonably question whether such groups can be well enough equipped and coordinated and last long enough to be able to participate seriously in any long-term observational effort such as might be required in the case of an arms control treaty. However, it certainly seems to be the case that interest in satellite spotting is increasing and the level of coordination and capability of the collective effort is also on the rise. This particular activity combines watching the sky, cataloguing difficult-to-spot objects, performing complex analyses, and communicating via the Internet. This has a long history of success in amateur circles and so is more likely to persist than not.

Question One: The Role of NGOs in Creating and Maintaining Momentum Toward an Agreement

Today, pressure from citizen groups can cause significant changes in government policy. They can also cause changes in the way and the speed with which government policy is implemented. The most recent example of this acceleration was seen in the case of the Ottawa Process which, in about one year, resulted in a multilateral agreement on banning the use of landmines. In a field in which multilateral arms negotiations can see a year go by with no progress whatsoever and in which decades of negotiation can pass before agreement is reached, this acceleration is, on the face of it, staggering.

Getting Accelerated Action

There may be reason to be optimistic about the prospects for a groundswell of support for worthy space initiatives. After all, space and virtually everything to do with it has tremendous resonance with many people. The recent experience with the Mars probe, the reporting of which grabbed headlines and led news broadcasts for over a week, demonstrates that much. Reaching back almost two decades, over one million people drove to the California desert to watch the landing of the first Shuttle from orbit. Prior to that, of course, was the legendary Apollo moon programme, whose descendant, the International Space Station, has been funded at several billion dollars per year for the past decade by a consortium of space-faring nations.

One could conclude that anything of historic significance to "space," therefore, would fairly easily lead to widespread interest among the public. It is also true that several high-profile groups of scientists have as part of their mandate to militate for arms control and even for disarmament. The Union of Concerned Scientists is a notable group in this category. Organizations primarily interested in "space" also express interest in moral and ethical issues through conferences and discussion groups. The Planetary Society, perhaps the best-known such organization, has in the past listed such events on its web site (www.planetary.org).

However, when it comes to the issue of whether or not to ban weapons for use against spacecraft or which are stationed in space for use against spacecraft, ballistic missiles, or targets nearer the Earth, one can imagine that synthesizing an effective, common viewpoint would be difficult. Although it should be possible to see that a treaty to ban the stationing of all weapons in space is very much in the long-term interest of humanity, no such initiative has so far captured the imagination of more than a handful of "space" adherents. Some of the reasons for this become clear when viewed against the contrasting experience of the landmine process so recently concluded in Ottawa. These include:

Emotional Resonance. No words are necessary to convey the suffering of innocent children caused by landmines left over from military conflicts. There is an immediate connection which transcends economic class, race, culture, or political persuasion. This emotional resonance made it possible for the landmines campaign to recruit the late Princess of Wales to the cause, whose own subsequent death made the issue even more poignant.

In contrast, arms control with respect to space has far fewer resonances. The most powerful idea motivating space enthusiasts is the idea of the exploration and conquest of new worlds. Many people can and do get excited about the possibility of people living on the moon or on Mars. Many others feel that it is inevitable that humanity will one day migrate to the stars. However, there is currently no clear-cut connection between these millennial dreams and the idea that weapons should form no part of the journey. Much to the contrary, the vision of space exploration presented in science fiction over the past fifty years, and most particularly since the end of the Cold War, has concentrated on weapons as an integral part of the quest. Thus, while we may share a vision in which the lives and limbs of children must be secured from the menace of landmines, the non-weaponization of space is, in comparison, a complex moral and analytical objective, but one not easily captured in a single, powerful image.

Military Neutrality or Support. Many people who supported the landmines ban would almost certainly not have done so if military experts had not stated clearly that landmines play no useful role. There is no such consensus on the role of weapons in space. Given the importance of space assets in both civilian and military operations, it is inconceivable that several categories of weapons intended to disable these assets do not exist in readiness today. The United States has clearly articulated its intention to control its own operations in space and to be ready to deny the benefits of space to any enemy during a time of conflict. While other space powers have not been as explicit, it must be assumed that this policy is not unique in the world. Since there is no significant agreement on how to assure the security of operations in space without weapons, in many countries the governing view of how to do so is, necessarily, the military one.

Moreover, it is not clear that all space enthusiasts see the non-weaponization of space as a desirable objective. The desire of many who seek the expansion of human operations into space is so great that the idea of "industrialization through weaponization" would not be abhorrent. That said, it should also be recognized that there are peace groups and other NGOs focussing on the interactions of technology with society which would readily align themselves with a project aimed at maintaining the non-weaponization of space.

Time Pressure. Space does not have weapons stationed in it at present. Although research and development on space-based weapons has certainly been under way in the United States and perhaps in other countries as well, it is unlikely that a political decision to deploy such systems is imminent in any of the few nations which would be capable of doing so. Even if space were littered with weapons, people would not be confronted with the same degree of imminent danger as is posed by landmines. Even if a large number of satellites were destroyed by a rogue weapon in space, few, if any, deaths would be caused directly by the withdrawal of satellite services. There would certainly be inconveniences, but, on the scale of major events, it would not rate above a large hurricane or medium-sized earthquake in its direct effects on populations.

However, this situation will not last. In the next decade, satellites will come to serve an increasingly important role in supplying infrastructure services in the developing world. As well, technology proliferation will admit more countries into the 'club' of space powers, and those countries which have that status today will become more capable and perhaps also more interested in anti-satellite weapons than they are at present. By the time a country is known to be developing space weapons, the thinking about whether such weapons are a good or bad investment will already be complete, resources will be committed, doctrines will be modified, and the time for arms control will be drawing to a close. Thus, although there is no pressing demand for space arms control today, there is a window of opportunity in which it could be accomplished.

Question One: Summary

In summary, then, a plan to involve public support and pressure to create momentum toward successful arms control in outer space must proceed on quite different lines from those pursued by the Ottawa Process. The key factors — emotional resonance, military acquiescence or support, and time pressure — are not compelling. However, a segment of space enthusiasts, used to having their eyes on a distant horizon, could perhaps support an initiative framed around long-term thinking which would allow them to feel in touch with a future which is otherwise out of reach. The question of what would ignite the passions of this group, and, once inflamed, how much it could accomplish is a difficult one to answer. It may be worthwhile to formulate a millennial proposal which would make explicit a vision of space without weapons, and to float this proposal through non-governmental channels to the community of space enthusiasts in order to see what their initial reactions would be. After all, although no significant groundswell of support has emerged for past attempts at space arms control, the reverse policy — the creation of weapons for use in space — has similarly not caught the public imagination in the few nations which could contemplate their creation.

Question Two: The Role of NGOs in Monitoring Compliance

It is not usual to consider the role of non-traditional observation and analysis in the verification of compliance with arms control agreements. However, space is unique and offers unique circumstances. Principally, space is visible for any observer on Earth. As a result, objects in space are not as anonymous as objects hidden on the ground. Whereas observers on the ground, or even in the air, can be kept beyond the horizon from land-based weapon systems, the laws of physics make this impossible to do for space-based weapon systems.

The skies are watched every night in every part of the world by professional and amateur astronomers as well as amateur satellite spotters. Consequently, there is always the possibility that a space-based object will be spotted and its orbital elements computed and published by a non-governmental observer somewhere in the world. It is essential to map out the role that such observations and such individuals would play in any arms control regime if embarrassment and misunderstanding is to be minimized or avoided. Because of their interest in space, these groups would invariably participate in the compliance monitoring process in some way. Given the rapid advance of technology, the technical assets for space monitoring which could be in NGO hands in the next decade could well rival those of the militaries of today. A compliance monitoring regime would be far more effective if it could channel NGO efforts, should these efforts prove to be significant.

Assuming that a space arms control agreement were concluded, it could deal with one or more of the following categories of weapons:

- Earth-based weapons (based and operated from on land, on and beneath the sea, or in the air); and,
- space-based weapons.

It could also deal with different aspects of the space weapon life cycle:

- research and development (R&D);
- testing;
- production and deployment;
- use; and,
- disposal.

The role of NGO observers is different for each of these categories of weapons and stages of the life cycle, as indicated in Table One and elaborated below. Again, the NGOs discussed here comprise technically aware professionals or hobbyists who either watch the skies or scan the literature for space-related news and who are in a position to synthesize views on the presence or absence of space weapons.

Table One: The Role of NGO Observers

	Earth-Based Weapons	Space-Based Weapons
R&D		
Testing		^
Production and Deployment		Potential Role
Use	***Limited Role***	***Limited Role***
Dismantling		Potential Role

When discussing compliance monitoring, it is often useful to place oneself in the role of a weapons "hider" or a weapons "finder" to see which role appears to be simpler to carry out.

R&D and Testing

Although it is difficult to generalize, one can argue that it is highly doubtful that NGO observers could play a role in detecting clandestine R&D or testing of space-based weapons. The reason for this is that these activities can be of very short duration and can be planned well in advance. Primarily, they can be timed to take maximum advantage of cloud cover and other events which block observation by random observers. Some material is published in open sources regarding research and development of weapon systems. It could be argued that investigation of open-source material could be a method of compliance monitoring. In the case of research into weapons intended to serve a deterrent role, this may be the case. After all, a weapon whose existence is not known to the enemy cannot serve any deterrent role. However, it is not clear at this time that space weapons would fit into this category. As a result, observation of open-source publications is not as good a tool for compliance monitoring as direct observation. In the case of R&D and testing, it is probably easier to be a "hider" than a "finder."

Deployment of Earth-Based Weapons

Monitoring deployment of earth-based weapons also admits very little role for NGO observers. The reason for this is that Earth-based weapons would presumably make only rare appearances in the open skies and would spend most of their time well hidden from any accidental exposure. Of course, once a weapon is used, then depending on the type of weapon it is, traces of its activities could well be observable long after the incident has taken place. Again, being a "hider" appears easier than being a "finder." Thus, the monitoring of the non-use of earth-based weapons is a marginal activity for NGOs.

Incidentally, of all weapon types directed against spacecraft, there are several persuasive reasons why this category makes the most military sense. First, weapons on the ground are less vulnerable to detection, evaluation, and countermeasures than weapons in space. Second, in most cases, they can be launched more quickly into the right orbit than can space-based weapons and thus are more effective at creating surprise. Third, experience in the Gulf War demonstrated that mobile ground-based systems are difficult to detect and hence relatively secure against counterattack. In comparison, a space-based weapon is difficult, if not impossible, to keep hidden once it is used. It then becomes vulnerable to counterattack. The capabilities of space weapons — such as lasers — to defend themselves against attack is open to debate. Certainly, their resources would be wasted in such a task if it meant that their primary mission would be compromised.

Deployment of Space-Based Weapons

The best potential for NGO monitoring is afforded by a regime which would ban the deployment of space-based weapons. It could be argued that for space-based weapons systems to be effective, they would either have to be comprised of many small weapon platforms, or a small number of large platforms. Either case heightens the potential of accidental discovery well before the system is used. It is possible, in theory, to design spacecraft to be invisible to the naked eye. However, designing to make them invisible to the growing array of sophisticated astronomical instruments would be highly problematic. And few things would be as embarrassing as deploying a clandestine system only to have it detected a few years later by instruments that were not even conceived of when the weapons were deployed.

Accordingly, the best way to hide space weapons would be in plain sight, disguised as ordinary satellites. Here, the persistence and enthusiasm of amateur satellite spotters would open a window of uncertainty which would be difficult for military planners to cope with. The art of satellite observation has progressed significantly since the days of the Kettering Group and is progressing still.

Satellite orbits are estimated and published independently of governmental agencies by amateur satellite watchers linked by the Internet. Systematic discrepancies or omissions involving a number of satellites would raise questions. Orbits at variance with the stated missions of spacecraft would also cause speculation and draw attention. Unusual maneuvers, appearances, or disappearances of spacecraft, and untypical radio transmissions would all galvanize the interest and efforts of amateur observers. At this point, it can be argued that the "finder" has an easier job than a "hider."

Use of Weapons in Space

There are various modes in which space weapons could operate. These modes are critical to consider when attempting to monitor the use of such weapons. Attack modes for weapons against spacecraft include the following major groupings:

- directed energy in various bands (IR, Optical, RF, etc.);
- undirected energy (nuclear detonation);
- electronic counter-measures — ECM (jamming, dazzling, spoofing, etc.); and,
- kinetic kill (physical contact, including debris from chemical explosions, causing damage).

Of these, the explosion of a nuclear weapon would be the easiest to detect. The reason for this is that upper atmospheric tests in the past demonstrated that the effects of an undirected nuclear weapon would affect the operations of a great many spacecraft, not just the ones passing through the area of the detonation. The whole near-Earth space environment would be highly perturbed, causing significant incidental spacecraft damage and blackouts of both in-orbit and terrestrial systems.

Of the rest, the weapons which act physically (kinetic kill weapons) are the most likely to be discovered. This is because a collision between such a weapon and a spacecraft would likely cause debris visible to observers, would change orbital elements in a characteristic way, would leave weapon remnants orbiting visibly, or would cause characteristic variations in the signals from the spacecraft's various operating sensors before the spacecraft was destroyed.

The use of directed energy weapons could be much more difficult to detect. The attack could take place in a relatively short period of time, and could be designed to take place when the Earth underneath the spacecraft was in daylight hours. This would minimize both detectability and the number of observers active. Weapons radiating either directly or incidentally in commonly monitored radiation bands could be detected. For example, an RF-based weapon would likely produce incidental radiation in other frequency bands and other directions than those associated with its weapon functions. Also, depending on the design of the weapon platform itself, changes in its own observable properties (thermal radiation or mass, for example) could persist over several hours, and therefore be detectable in principal by ground-based observers. Space-based lasers could well vent characteristic and observable chemicals. In the extreme case of the X-ray lasers discussed as part of the Strategic Defense Initiative (SDI) programme in the 1980s, the satellite would be destroyed as part of its own operation.

Finally, the operation of weapons in an Electronic Counter-Measures (ECM) mode would be difficult indeed to observe; especially in a spoofing mode where signals are sent to the target which cause the target to become ineffective in its role. It is not easy to operate a spacecraft at the best of times, and pinning failures on a hostile agency could be difficult to do even for the operator of the spacecraft. The jamming of a spacecraft or the hijacking of its uplink signal would, of course, be much easier to detect for the spacecraft operator. However, this detection would be practically impossible for independent NGO observers.

In summary, then, there is reason to believe that the use of certain types of weapons in space could be observed by NGO "finders." However, their role could be limited or even unfeasible, depending upon the type of weapon system used by the "hiders."

Dismantling of Space Weapon Systems

Finally, NGO observers could witness the dismantling of the space-based portion of a weapon system. Assuming that these were deployed and withdrawn again with a subsequent treaty, NGOs could monitor a list of declared spacecraft and track their reentry and destruction in the Earth's atmosphere.

General Issues Related to the Effectiveness of NGO Observers

Developments in launch system technology are making observation of spacecraft more difficult. Whereas during most of the Cold War, military launches would fairly consistently come from a small number of installations, launcher technology is quickly expanding the range of possibilities. It is more difficult to detect satellites when the launch system can be airborne or seaborne. Added to this is progress on direct-to-orbit launchers which can, in principle, take off and land like aircraft. When there is significant uncertainty about the location and timing of a spacecraft launch it is more difficult for amateur observers to detect the spacecraft.

Although space is uniquely visible from the ground, it is also extremely vast. Satellites have been "lost," for instance, when an onboard explosion caused the satellite orbit to change quickly. Satellites in orbit near the Earth are more likely to be spotted even if their existence is not previously known. Also, their brightness and rapid motion relative to the stars make amateur observations easier. However, for satellites operating at higher orbits, losing a significant reference point such as the launch site makes it more difficult to detect the satellite while it is in a transfer orbit, and thus makes its detection more difficult overall.

The number of satellites in orbit is expected to grow significantly as well. This will create a great deal of background activity and make more room for clandestine satellite operations. It will also make satellite observation a more interesting hobby and thus focus more interest on space. It is not clear whether these new space operations will make it easier to determine what is "routine" and what is "unusual," or whether it will make it likely that a wider array of activities will make it difficult to typify any behavior as "unusual."

The proliferation of computer communications capabilities in the 1980s and 1990s has made it much easier to coordinate the activities of a widely dispersed group of people. This was true even before the Internet expanded to its present capabilities. Interested people could sign on to private bulletin board systems to download satellite orbital information and to exchange tips and techniques. With the Internet's World Wide Web, attracting more interested users is easier and the amateur satellite observation community is growing.[1]

This use of a fully open communications system (the Internet) leaves open the possibility that the system could be subverted with disinformation. Various incidents of Internet-based rumor-mongering have been observed in the past. If country X wanted to create problems for country Y, it could spread rumors about Y's satellites, thus focusing attention and perhaps even generating significant publicity. Although this tactic would undoubtedly work in diverting some individuals, there is no guarantee that everyone would be fooled, so that the overall effect would be weakened. There is also a small but significant chance that the tactic would backfire causing even more problems for country X.

Alternatively, parties interested in disabling the monitoring effort for a period of time could launch various types of "denial of service" attacks which would disable the site. Done once, this would leave a window of opportunity in which clandestine space activities could take place. However, such a manouevre would be fairly obvious and would probably cause observers to pay more attention to see what might have been happening. Done periodically, this would lower the confidence of the participants in their communications system and make them set up a more secure one. Either way, the potential exists that any direct attack on the communal communications infrastructure could backfire.

In the past few years, orbital dynamics software, as well as computers powerful enough to run sophisticated orbit propagation calculations, have become widely available. As well, the knowledge of how to use these systems is widely available in the persons of well-educated, enthusiastic amateurs easily accessible through the Internet. This combination of factors means that the "entry barriers" to success in amateur satellite spotting have lowered dramatically.

---

1. One important such web site can be found at http://www.satellite.eu.org/sat/vsohp/satintro.html. This web site is reported to get 15,000 hits per month. An associated mailing list has over 450 subscribers. This community has applied for its own Usenet newsgroup (under sci.astro.satellites.visual-observe), and voting is to take place in March 1998 to see whether this new newsgroup will be accepted.

Finally, it is not inconceivable that amateur groups could in the next decade launch space telescopes of their own. For example, AMSAT, the amateur satellite organization that has launched numerous spacecraft over the last few decades, is now negotiating with a Canadian contractor on a joint project to make a space telescope microsat largely under Canadian Space Agency funding. This space telescope microsat has a well defined scientific mission, but if it survives long enough in space, it could well be configured to make observations on behalf of users anywhere in the world.

Question Two: Summary

Even though space is visible from anywhere on Earth and even though satellites can be and are routinely spotted by amateur observers with the naked eye or with widely available lenses, the participation of NGOs in the monitoring of compliance with a space arms agreement would necessarily be limited.

The most compelling case which can be made from among all the possibilities analyzed herein, is that NGOs could and would participate in the monitoring of a ban on space-based weapons. In this scenario, the restrictions on their participation could perhaps be offset by their enthusiasm, persistence, and ingenuity. Some framework to capture this participation would be highly recommended.

Part II
Extending the Agenda

Chapter Twelve

A View on the Application of Overhead Imagery to Strengthen the IAEA Safeguards Regime

Hui Zhang*

Abstract

An important way to prevent the proliferation of nuclear weapons is to control the production and use of fissile materials. The main international controls on fissile materials are embodied in the Non-Proliferation Treaty (NPT). The NPT requires that the countries joining it as non-nuclear weapon states accept "full-scope" IAEA safeguards. However, the traditional IAEA safeguards system cannot effectively detect undeclared activities. Moreover, it can be expected that the next important measure to control the production of fissile material will be the cutoff convention. This cutoff convention also needs to develop techniques to detect undeclared nuclear activities. To strengthen IAEA safeguards and verify the fissile cutoff convention, it is necessary to apply and develop remote sensing techniques to detect covert nuclear facilities.

Given the dramatic increase in overhead imagery capabilities — especially commercial observation satellites — more and more people are beginning to appreciate the role of overhead imagery in arms control, confidence building, and peacekeeping. The purpose of this chapter is to examine the possibilities for applying present and projected overhead imagery — particularly commercial observation satellites — to strengthen the IAEA safeguards regime. First, I briefly review the present and projected overhead imaging systems which are applicable. Second, I discuss the synergistic aspects of the

---

* This chapter is an edited version of a presentation made to the Fifteenth Annual Ottawa NACD Verification Symposium, "Arms Control and the Rule of Law: A Framework for Peace and Security in Outer Space," 11–14 March 1998, Montebello, Quebec, Canada. The author would like to acknowledge the encouragement and support of the Non-Proliferation, Arms Control and Disarmament Division of the Department of Foreign Affairs and International Trade, Canada. The author would also like to thank Dr. James Keeley of the University of Calgary for his helpful comments. Finally, the author extends thanks to his colleagues at the Center for Energy and Environmental Studies at Princeton University for helpful discussions. Of course, the author assumes full responsibility for this study and its conclusions.

use of such systems with existing methods of monitoring and verification under the IAEA safeguards programme. Third, the utility of applying such systems to the monitoring and verification of a possible fissile material cutoff treaty is discussed. Finally, some suggestions are made regarding the appropriate organization and management of such a system under the IAEA.

Introduction

The proliferation of nuclear weapons continues to be an important international security issue. How to strengthen the non-proliferation regime is a central question. One way to prevent the proliferation of nuclear weapons is to control the production and use of fissile materials (plutonium and highly enriched uranium). The main international controls on fissile materials are embodied in the NPT. The Non-Proliferation Treaty requires that countries joining it as non-nuclear weapon states accept "full-scope" IAEA safeguards. However, the traditional IAEA safeguards system cannot effectively detect undeclared activities at safeguarded sites or facilities at undeclared sites. Moreover, it can be expected that the next important measure to control the production of fissile material will be a cutoff convention, which would ban the production of fissile materials for weapons.[1] Such a cutoff convention will also need to develop techniques to detect undeclared nuclear activities. To strengthen its safeguards, the IAEA is developing and researching environmental monitoring measures for both on-site and off-site verification.[2] However, in the case of clandestine nuclear sites, it is very difficult to conduct on-site verification because it is usually not known where the site is, or the country in question may severely restrict access to its nuclear sites and facilities under some pretext in order to hide its undeclared production activities. Thus, it is very important to apply and develop remote sensing techniques in order to detect covert nuclear facilities. Given the dramatic increase in overhead imagery capabilities — especially commercial observation satellites — more and more people are beginning to appreciate the role overhead imagery might play in arms control, confidence building, and peacekeeping.[3]

---

1. See F. Berkhout, et al., "A Cutoff in the Production of Fissile Material," *International Security*, 19:3 (Winter 1994/95).
2. United States, Congress, Office of Technology Assessment, *Environmental Monitoring for Nuclear Safeguards* (Washington: U.S. Government Printing Office, 1995).
3. See, for example, Michael Krepon, et al., *Commercial Observation Satellites and International Security* (New York: St. Martin's Press, 1990); Michael Slack and Heather Chestnutt, eds., *Open Skies: Technical, Organizational, Operational, Legal and Political Aspects* (Toronto: Centre for International and Strategic Studies, 1990); Michael Krepon and Amy E. Smithson, eds., *Open Skies, Arms Control, and Cooperative Security* (New York: St. Martin's Press, 1992).

Applicable Present and Projected Overhead Imaging Systems

Technical Considerations

For detecting clandestine nuclear activities, certain technical capabilities with regard to overhead imaging are required.

Spatial resolution is defined as the minimal distance two identical objects must be separable in the image. The quality of image data from overhead imaging systems is mainly characterized by their spatial resolution. High-resolution imaging is required for detection, identification, and description of various objects and activities. Table 1 gives a typical estimate of the ground thresholds for interpreting various targets. It is generally agreed that 3-5 resolution elements of pixels are required to recognize a detected feature and up to 10 elements are required to describe a feature or object.[4] By comparison with these values, the monitoring requirements for various nuclear activities and facilities can be estimated. The resolution obtainable by an imaging sensor depends not only on the characteristics of the detector itself and of the optical system, but also the orbit height, the wavelength of the radiation to be detected, and the contrast ratios of the target object to its background.[5] In addition, spectral resolution across a range of wavelengths is useful for sorting out areas of vastly different composition but similar reflectance. Consequently, if the scene characteristics are favourable to overhead observation, objects can be detected and identified at lower spatial resolutions than those listed in Table 1.[6]

Revisit time is the minimum time that must elapse before a single imaging satellite can return to a specific geographic point. Revisit time is dependent upon sensor (geographic coverage by optical field of view or pointability) and the flight parameters of the sensor platform.[7] For timely detection of clandestine nuclear activities, a shorter revisit time is favourable. For this purpose, a sensor system with variable viewing angles is usually required. Pointing to the left and right of the satellite can significantly decrease the revisit time. Moreover, oblique imaging can be used to acquire stereo pairs. It is preferable, however, to acquire a stereo pair in a single orbital pass through fore and aft viewing. With this stereo imaging, objects of interest will be easier to identify.

---

4. R. Ondrejka, "Imaging Technology," *Arms Control Verification: The Technologies That Make it Possible*, Kosta Tsipis, et al., eds. (Washington: Pergamon-Brassey's International Defense Publishers, 1986).
5. Thomas M. Lillesand and Ralph Kiefer, *Remote Sensing and Image Interpretation* (New York: John Wiley & Sons, 1994).
6. Peter D. Zimmerman, "Verification With Civilian Remote Sensing Satellites: An Introduction to the Technology and a Review of the Events of 1989 and 1990," in *Verification Report 1991*, J.B. Poole, ed. (London: VERTIC, 1991).
7. P.R.K. Chetty, *Satellite Technology and Its Applications* (Blue Ridge Summit: TAB Professional and Reference Books, 1991).

Table 1. Ground Resolution in Metres[8]

Target		Detection	General ID[b]	Precise ID[c]	Description[d]	Technical Analysis
Bridges		6	4.5	1.5	1	0.3
Communications:	Radar	3	1	0.3	0.15	0.015
	Radio	3	1.5	0.3	0.15	0.015
Supply Dumps		1.5–3	0.6	0.3	0.03	0.03
Troop Units (in bivouac or on road)		6	2	1.2	0.3	0.15
Airfield Facilities		6	4.5	3	0.3	0.15
Rockets and Artillery		1	0.6	0.15	0.05	0.045
Aircraft		4.5	1.5	1	0.15	0.045
Command and Control HQ		3	1.5	1	0.15	0.09
Missile Sites (SSM/SAM)		3	1.5	0.6	0.3	0.045
Surface Ships		7.5–15	4.5	0.6	0.3	0.045
Nuclear Weapons Components		2.5	1.5	0.3	0.03	0.015
Vehicles		1.5	0.6	0.3	0.06	0.045
Land Minefields		3–9	6	1	0.03	0.09
Ports and Harbours		30	15	6	3	0.3
Coasts, Landing Beaches		15–30	4.5	3	1.5	0.4
Railroad Yards and Shops		15–30	15	6	1.5	0.15
Roads		6–9	6	1.8	0.6	0.4
Urban Areas		60	30	3	3	0.75
Terrain		—	90	4.5	1.5	0.75
Surfaced Submarines		7.5–30	4.5–6	1.5	1	0.03

a. Detection: location of a class of units, object, or activity of military interest.
b. General identification: determination of general target type.
c. Precise identification: discrimination within target type of known types.
d. Description: size/dimension, configuration/layout, components construction, equipment count, etc.
e. Technical analysis: detailed analysis of specific equipment.

8. Table reproduced from Ann M. Florini, "The Opening Skies: Third-Party Imaging Satellites and U.S. Security," *International Security*, 13:2 (Fall 1988), 98.

Observation Satellites

Sometime after the Soviet Union and the United States launched their first satellites (Sputnik-1 in October 1957 and Discover-1 in February 1959, respectively), military reconnaissance satellites became important national technical means (NTM) to verify some arms control agreements such as SALT I (1972), the ABM Treaty (1972), and START I (1994). As satellite images from the US and Soviet classified domains improved from the 1960s to the 1980s, commercial observation satellites also began to develop rapidly, starting with the launch of US Landsat-1 in 1972 — whose spatial resolution in the visible spectrum is 80m.[9]

In 1986, France launched SPOT-1, whose spatial resolution is 10m. Its revisit time is one to four days.[10] It was with these moderately high-resolution (10m) images that many experts and news media started to pay more attention to the arms control and military significance of these commercial images. In 1987, Soviet KVR-1000 imagery,[11] whose spatial resolution can achieve approximately 5m, appeared on the market. It was then that the competition for high-resolution commercial satellite imagery began. France's SPOT-2 was launched in 1990; SPOT-3 was launched in 1993, but as a result of an incident on 14 November 1997 it has stopped functioning; SPOT-4 is scheduled for launch in early 1998. Currently, SPOT-1 and SPOT-2 are in operation. All these SPOT satellites have a resolution of 10m for the panchronic mode. SPOT-5, scheduled to be launched in 2002, will have a resolution of 3m. Recently, Russia began to provide 2m resolution images — the highest resolution images to this point. However, since its images are acquired on photographic film which is delivered back to Earth inside a reentry vehicle, customers must usually wait a considerable time to get their images. Moreover, its mission is relatively short (a typical lifetime is 28 days) and differs from other nations' satellite systems in that the satellites are not in sun-synchronous near-polar orbit. Thus, this satellite system is not suitable to monitoring global nuclear activities with regularity.

The most important development to date in the field of high-resolution commercial satellites could be the launch of the US EarlyBird on 24 December 1997. Its spatial resolution is 3m. On 9 March 1994, the White House approved a radical policy change that allowed US companies to build and launch private observation satellites that would return imagery with resolution of 1m or better.[12] Consequently, it can be expected that by the end of 1998, commercial satellite image resolution at 1 metre or better will be available.

---

9. *Op cit.*, Lillesand and Kiefer.

10. *Ibid.*

11. Vipin Gupta and Philip McNab, "Sleuthing from home," *Bulletin of the Atomic Scientists*, 49:10 (December 1993).

12. United States, White House, "Fact sheet on U.S. policy on foreign access to remote sensing space capabilities," 10 March 1994, 1.

Therefore, it is appropriate to turn our focus here to these new satellites with spatial resolution of less than 3m (see Table 2).

Commercial Observation Satellites

EarlyBird. This satellite was launched by the US firm EarthWatch on 24 December 1997, and is designed to operate for five years.[13] The EarlyBird has a circular, near-polar sun-synchronous orbit. The satellite system carries a 3m resolution panchromatic sensor and a 15m resolution multispectral sensor. Its average revisit time is two to five days depending on the latitude. It can expose single images as small as 9 square kilometres or up to 900 square kilometres at a time. Thirty degree fore-aft and 28 degree side-to-side tracking enable collection of images over a broad area, as well as in-track stereo images. *QuickBird* will also be launched by EarthWatch in late 1998.[14] It will carry more advanced sensors onboard: 0.82m resolution panchromatic sensor and a 3.28m resolution multispectral sensor. Its average revisit time will be one to four days. Its image sizes will range from a 484 square kilometre snapshot to a 40,500 square kilometre strip map. Its multispectral band will include blue, green, red, and near-infrared, so that it can get true color images from spaces. It has 30 degree fore-and-aft and side-to-side pointing capability, as well as stereo images. All these EarthWatch satellites have a huge capacity for onboard data storage. Using advanced satellite technology, EarthWatch can delivery its imagery to customers within hours to days, instead of weeks or months as before.

The year 1998 will be an important year for high-resolution commercial observation satellites. Other satellites, whose spatial resolution can arrive at 1m, will also soon be in orbit (see Table 2). The US firm Space Imaging will launch its Space Imaging Satellite (SIS) in the spring of 1998.[15] The US company Orbital Imaging will launch the OrbView satellite this year. In mid-1998, Israel will launch the Earth Resource Observation Satellite (EROS-A), followed by EROS-B in 1999. All these satellite systems will have similar features as those of the QuickBird satellite: high spatial resolution, shorter revisit time, etc. At the same time, some other countries also have had or soon will have high-resolution observation satellites: Canada launched RADARSAT in 1995 (about 9m resolution); India's IRS-1C and IRS-1D (scheduled for 1999, with 10m resolution); and South Korea's Komsat (scheduled for 1998, with 10m resolution). It can be expected that as a result of market competition, more and more satellites with 1m or better resolution will be in orbit.

It should be noted that the sensors on board these satellites, getting a resolution of 1m, operate on the spectrum of visible light and near infrared (VNIR). These fine resolution images could be very useful for identifying

13. See http://www.digitalglobe.com/company/satellites.html.

14. *Ibid.*

15. See http://www.ccrs.nrcan.gc.ca/ccrs/tekrd/satsens/sats/asprse.html.

nuclear sites and facilities by their physical features. However, this kind of imaging can mainly acquire data in daytime and under favourable weather conditions (i.e., without clouds). For viewing under other conditions, thermal infrared sensors (TIR) and microwave radar systems (such as SAR) would also be required.

Table 2: Present and Projected Commercial Satellites with Fine Resolution

Satellite System	Country	Schedule Date	Spatial Resolution (in metres)	Revisit Time[a] (Days)	Maximum Viewing Angle (in degrees)	Orbital Attitude (in km)
EarlyBird	US	Dec. 1997	3	4.75	30	470
QuickBird	US	mid-1998	0.82	2	30	600
SIS	US	1998	1	2	30	680
OrbView	US	1998	1	3	45	460
EROS-A	Israel	mid-1998	2	3	30	480
EROS-B	Israel	1999	1	3	30	480

a. Revisit time at equatorial latitudes.

TIR. One of the important signatures of some nuclear activities is their heat output. The TIR is sensitive to this heat radiation.[16] In contrast to sensors for VNIR, TIR can work in both daytime and nighttime. The spatial resolution for TIR is similar to the values listed in Table 1. However, even smaller objects could be detected for hot spots when the thermal difference between object and background is high enough. It can also detect narrow and long hot objects. Usually the best TIR imagery will have a resolution about 10 to 100 times poorer than the best visible-light photography. Besides the spatial resolution, the temperature sensitivity of TIR is needed. Objects with temperatures below 100°C emit dominantly at wavelengths above 8μm, so here we mainly conisder the 8–14μm (within the atmospheric window) band.

The existing and projected satellites with TIR sensors (8–14μm spectrum) are given in Table 3.[17] The capability of existing or soon-available TIR is a spatial resolution of 40m and a temperature sensitivity of 1°K.

---

16. R. Reimers, *The Technological Aspects of Remote Sensing for Verification Purposes*, 1989.
17. Source: http: //www.asterweb.jp.nasa/gov/asterhome; a chart from Vipin Gupta.

Table 3. The Satellites with TIR Sensors (8-14μm spectrum)

Satellites	Scheduled date	Spatial resolution (in metres)	Temperature sensitivity (°K)
Landsat TM	1982	120	8
Landsat ETM+	1998	60	8
Aster	1998	90	1
MTI	1999	40	1

SAR. Unlike VNIR and TIR sensors, radar systems can operate under any weather conditions (i.e., cloud cover, smog, and haze), as well as in daytime and at night. However, because of the long wavelength, real antenna radar has worse spatial resolution. Images with good resolution can only be obtained using Synthetic Aperture Radar (SAR). An important development in this respect was the launch of the Canadian RADARSAT satellite in November 1995.[18] RADARSAT has the unique ability to shape and steer its radar beam over a 500-km range. Its beams can be turned from the ground. They can image swath from 35-km to 500-km with varied resolution. For its fine resolution mode, the range resolution and azimuth resolution can be 9m. Russia could launch its Almaz satellite with SAR, whose resolution could arrive at 5m.

Military Observation Satellites

Until now, the satellites with the highest resolution have been monopolized by the US and Russia. Since November 1992, the US has launched the KH-12 series satellites.[19] The KH-12 sensor, operated in VNIR, can reach 10cm-resolution images (10cm resolution is at the limitation for space imaging because of atmospheric distortion[20]). It also carries TIR to detect heat sources, but its resolution is not known. Additionally, it can also take pictures at very high angles of obliquity and image objects hundreds of kilometres away from its flight path. It can digitally relay imagery to Earth in near real-time. In addition, since December 1988, the US has launched the Lacrosse series with SAR.[21] Its resolution could be 1.3m to 3m. The KH-12 and Lacrosse satellite systems will give the US a capability for real-time, all-weather, and high-resolution imaging. France's Helios military satellite project will have 1m or better resolution.

18. See http://www.ccrs.nrcan.gc.ca/ccrs/tekrd/radarsat.
19. See http://www.fas.org/spp/military/program/imint/kh-12.html.
20. Johan Swahn, *Open Skies for All: the Prospects for International Satellite Surveillance*, 1989.
21. *Ibid.*, 56.

Aerial Observation

To strengthen the safeguards system and to verify the cutoff treaty, some aerial observation could be specified. Practically, many types of existing aircraft can be used as remote-sensing aerial platforms.[22] The airborne platform could include fixed-wing aircraft, helicopters, and airships. All types of sensors discussed above — such as the VNIR, TIR, and SAR — can be used. In addition, to detect a nuclear fingerprint, other sensors could be included such as Lidar (light detection and ranging), which could determine the concentration and composition of the effluents discharged from the nuclear operation.[23] The imaging sensor can get very fine spatial resolution compared to that from a satellite, because of its much lower altitude. For example, when we use the 3m-resolution panchromatic sensor of EarlyBird, whose orbital altitude is 470 km, we can get about 10cm-resolution from a 15 km altitude aerial platform. However, unlike the satellite case (with no legal limitations), the type and accuracy of the sensor on the aerial platform and the height of the airborne platform will be restricted by political, military, and national security factors. Here we can refer to the Open Skies Treaty.[24] This is an important confidence building measure that allows personned aerial overflight of participating nations with multi-sensor suites in order to monitor any militarily significant activities. The Treaty constrains the sensors and the resolution specifications of those sensors. It permits:

- optical panoramic and framing cameras and video cameras with real-time display with resolution of 30cm;
- infrared line-scanning devices, with resolution of 50cm; and,
- sideways-looking synthetic aperture radar, with resolution of 3m.

The Treaty allows for upgrades of sensor capabilities and for the inclusion of other new sensor types in the future. An additional requirement of the Treaty is that all technology used by an observing party must be commercially available to all other signatories to the Treaty. In short, this Treaty provides an opportunity for the application of aerial observation to IAEA safeguards and a cutoff treaty.

---

22. F.R. Cleminson, "Aerial Monitoring for Verification Purposes," in *Unconventional Approaches to Conventional Arms Control Verification*, John Grin and Henny van der Graf, eds. (Amsterdam: VU University Press, 1990); Allen V. Banner, *Overhead Imaging for Verification and Peacekeeping: Three Studies* (Ottawa: External Affairs and International Trade Canada, 1991).
23. Raymond Measures, *Laser Remote Sensing* (New York: John Wiley & Sons, 1984).
24. J. Allentuck, "Open Skies: Facilitating the Many Dimensions of Transparency," INMM 34th Annual Meeting Proceedings, Vol. XXII, July 1993.

Synergistic Aspects of the Use of Such Systems with Existing Methods of Monitoring and Verification under the IAEA Safeguards Programme

The Detection of Clandestine Facilities using Overhead Imagery

The traditional safeguards method mainly focuses on declared nuclear facilities and activities. However, it cannot effectively detect undeclared facilities at a safeguards site or facilities at undeclared sites. Although the IAEA is promoting the "93+2" programme to strengthen safeguards for detection of such cases by environmental monitoring techniques,[25] it will still be very difficult to detect undeclared nuclear sites. Commercial observation satellites will play an important role in such cases.

Commercial Satellites and Verification

Satellites have the advantage of searching for and locating clandestine nuclear sites by surveying the whole globe with regular and frequent visits and with fine resolution.

Detecting undeclared nuclear sites. Undeclared sites could be detected by their infrastructure signatures through the satellite images. To produce fissile material for weapons, the key infrastructures (uranium enrichment facilities and/or production reactors and reprocessing plants) must be established. These plants could have some infrastructure signatures which can be detected and identified by satellite imagery with fine resolution (such as 1m). Figure 1, depicting the Dimona nuclear facility in Israel, illustrates this.[26] The photo in the left frame was taken by the US CORONA satellite, on mission 1115-2 on 29 September 1971. From the recent declassification of information from KH-1 to KH-6, which collected images of the Earth's surface between 1960 and 1972,[27] we find that this photo must have been taken by KH-4B. Its best ground resolution is 1.8m by panchromatic mode. The photo in the right frame was taken by SPOT (10m resolution). From these two photos, we can see that the physical structure of the nuclear complex of Dimona is very clear by the 1.8m resolution image. In fact, even by the 10m resolution SPOT image, it could be detected, although the picture is not as clear. If we use the QuickBird image of 0.82m resolution, the two pixel size will be about 1.6m, and it should be comparable with the KH-4B photo. Consequently, the 1m resolution satellite could detect most of the nuclear facilities.

---

25. S. Moyland, *The IAEA's program "93+2"*, January 1997.
26. See http://www.fas.org/irp/imint/dimona_compare.html.
27. See http://edcwww.cr.usgs.gov/glis/hyper/guide/disp.

Figure 1. Images from KH-4B and SPOT of the Dimona Nuclear Facility in Israel

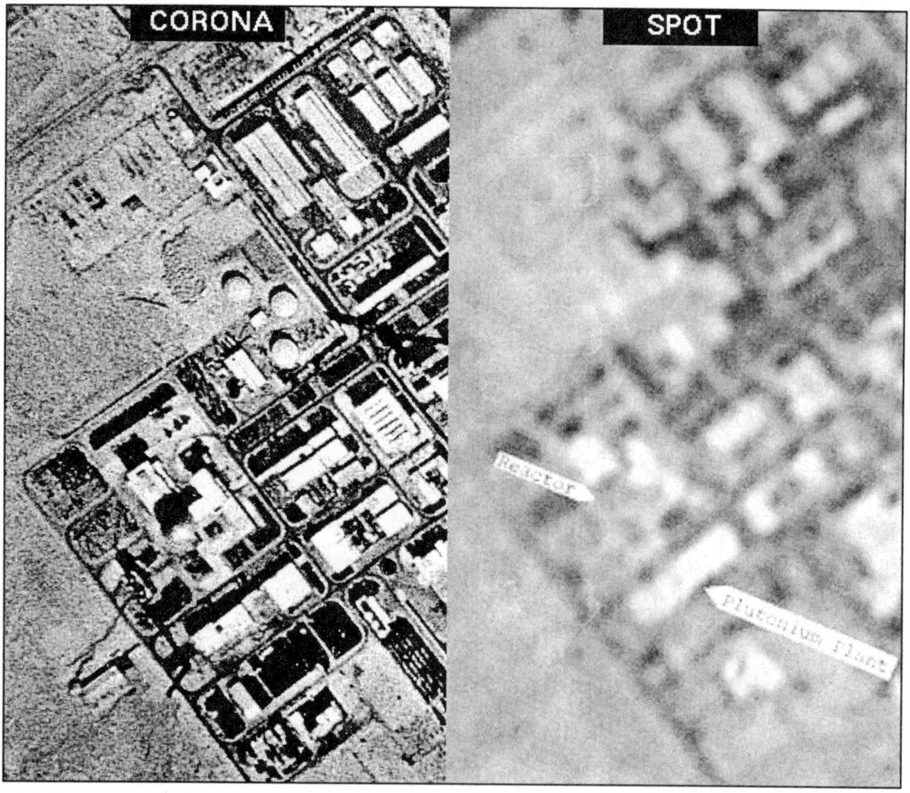

Comparison of Imagery from SPOT and CORONA missions:
1115-2 — 29 September 1971 — Frame: 52,53

In the case of a reactor, we can expect a large main reactor building and cooling towers or other water source. If the reactor uses cooling towers, it is usually the natural draft hyperbolic cooling tower.[28] For a 50-MW reactor that could produce two bombs worth of fissile material per year, the cooling tower size is usually of several tens of metres diameter and height. It should be very easy to detect such a large physical structure by 1m resolution images. When the reactors are operating, the water-vapor plumes from the cooling tower should be especially easy to see. Moreover, to produce plutonium, the operation time of the reactor is at least several months. Thus, a satellite with a revisit time of several days should be sufficient to detect the plume. Similarly,

28. W. Stanford, et al., *Cooling Towers: Principles and Practice* (Boston: Butterworths, 1972).

if the reactor uses a natural water body as the cooling system, the warm-water plumes from surface-water heat exchangers could be visible. For a reprocessing plant, its infrastructure usually exhibits a "canyon-like" construction (typical of large reprocessing plants), high stack, underground facilities, tanks for liquid waste, and buildings with thick walls. Moreover, its construction or operation will include a great many activities, such as the shipment of various materials. The construction period for a reprocessing plant or reactor is usually in the order of years. The operation of a reprocessing plant is at least several weeks. Therefore, satellites with several days revisit time and 1m resolution could detect these facilities and activities.

For gaseous diffusion uranium enrichment, the plants structural integrity and layout will include multiple large-area process buildings, storage yards for UF6 cylinders, nearby electrical generating facilities, and cooling towers or a nearby water body. Moreover, once the gaseous diffusion enrichment plant is started, its operating time is at least several months. As above, these facilities could be detected by space imaging with 1m spatial resolution and several days revisit time. Finally, all these sites are usually located away from cities and within security districts, which could provide targets for satellites searching areas.

The above discussions focus on VNIR imaging. In addition to this, during these facilities' operation, their discharged heat effluents could be detected by TIR. However, for the heat effluent from reactors or enrichment facilities, with the present capability of TIR on board the satellite (40m spatial resolution and 1°K temperature accuracy), it is difficult to detect small-scale production facilities, which are likely in the case of non-nuclear weapons states. However, in the case of a large production level, the heat effluents could be detected.

Detecting undeclared nuclear facilities at declared sites. A proliferator could build clandestine nuclear facilities inside or near a declared site. It should be easier to detect these facilities than those at unknown sites, because the satellite would target and observe these declared sites frequently. If some new undeclared buildings and facilities appear in these areas or some changes take place, the satellite can detect these activities by comparison with earlier photos. It could then trigger the IAEA to conduct on-site special inspections. For example, it was US satellite photos taken around the declared nuclear sites at Yongbyon, North Korea, that triggered the IAEA inspection of two suspected waste sites and one building (suspected to be used for reprocessing activities) around the radiochemical laboratory at Yongbyon.[29]

In summary, with the present visible and near infrared images (1m resolution) clandestine nuclear sites can be detected. Once the satellite locates a suspect site, a subsequent aerial observation flight might be directed for further investigation or even trigger an on-site challenge inspection.

---

29. David Albright, "North Korean Plutonium Production," *Science & Global Security*, 5:1 (1994).

Aerial Observation

In contrast to satellite imaging, aerial observation is not very effective in searching for clandestine nuclear sites. However, in the case of known suspect sites (source could be from the satellite, NTM, or human intelligence), aerial observation would be useful for confirmation. Most importantly, it should be easier to detect undeclared facilities around a declared site by monitoring changes. With the fine resolution images allowed for in the Open Skies Treaty, all the clandestine nuclear facilities and activities discussed above will be much easier to identify. Moreover, based on the allowed 50cm resolution of TIR, the heat effluent from a reactor or uranium enrichment facility should be very easy to detect.

However, for some sites, use of the very fine VNIR sensors or on-site inspection might not be permitted by some states for reasons of military or industrial security. Thus, it is necessary to develop remote sensing techniques that can only detect the nuclear characteristic effluent but which cannot see the physical structure. This could be achieved by using Differential Absorption Lidar (DIAL) to detect free Iodine-129 molecules released from the stack of the reprocessing plant. The discharged Iodine-129 can be detected at a distance of 1 km by using existing dye lasers.[30]

Detection Synergism

To strengthen IAEA safeguards, the safeguards system should include the existing safeguards, environmental monitoring techniques, and overhead imagery.

For declared nuclear facilities and activities, the IAEA should mainly apply the existing safeguards system, based on nuclear accountancy, independent verification of declared nuclear material — i.e., by item counting, measurement, and taking samples and analysis — and auditing of the books of the operator. The material accountancy has been used as a safeguards measure of fundamental importance. It is supplemented by "containment" and "surveillance" measures. Moreover, environmental monitoring techniques (one focus of the "93+2" programme) could be required. When nuclear facilities are in operation, some characteristic effluents will be released into the environment. Thus, the environmental sampling of air, water, or soil, on- or off-site would be useful to detect clandestine production activities. In the case of declared nuclear facilities and activities, overhead imagery would not be very useful.

In the case of detecting undeclared activities at a safeguarded site, the traditional safeguards system is not effective. Overhead imagery and environmental monitoring techniques will be useful. Although satellite images and aerial observation can find clandestine buildings or facilities, they cannot

---

30. Zhang Hui and Du xiangwan, "Remote Sensing of Undeclared Nuclear Reprocessing Plants by DIAL Techniques," The Eighth International Summer Symposium on Science and World Affairs, Beijing, July 1996.

make the final determination. However, they could trigger special inspections with on-site sampling analysis and visual observation.

Where an undeclared nuclear site is concerned, traditional IAEA safeguards and environmental monitoring techniques will not be very effective because it will be very difficult to conduct them without a target. However, satellite imaging could be an important method of detecting and locating these suspect sites. It can then direct aerial observation to further verify the sites, conduct off-site environmental monitoring at sites and, finally, even trigger a special inspection with on-site sampling analysis. In addition, NTM or human intelligence may provide information of clandestine activities for satellite sensing, aerial observation, environmental monitoring, and on-site special inspection. Consequently, an effective safeguards system including all these methods would be synergistic.

Applying such Systems to Monitor and Verify a Possible Fissile Material Cutoff Convention

A global, verified ban on the production of fissile material for nuclear explosives would be an essential part of any comprehensive nuclear disarmament and non-proliferation regime. The primary goal of a cutoff convention will be to attain the signatures of the five declared and three *de facto* nuclear weapon states (India, Israel, and Pakistan), because virtually all other countries are already subject to its requirements by virtue of their ratification of the NPT as non-nuclear weapon states. A principal focus in negotiating a cutoff treaty will be the verification provisions. Verification measures must be seen as effective and politically acceptable. The cutoff treaty's verification scope should include the facilities and activities which could be used for the production or the diversion of weapons-usable fissile material. For these nations' civilian nuclear power sectors, it should mainly apply the existing safeguards system and environment monitoring techniques. For undeclared nuclear activities, overhead imagery, environmental monitoring techniques, and the existing safeguards regime could be applied as discussed previously. Most important, verification should focus on past military nuclear production facilities. After the conclusion of a cutoff convention, some of these facilities could continue operating for civil nuclear power or non-weapon military uses. In such cases, overhead imagery would be less useful. However, most of these production facilities could be shut down. In these cases, overhead imagery could be readily used to monitor their status.

Production Reactors

As noted above, heat output is the most obvious characteristic effluent of an operational reactor. For example, a warm water-vapor plume may be discharged from the cooling tower exits. These vapor plumes should be easy to see with the 1m-resolution images onboard satellites. For the once-through cooling system, it has been found that the discharged hot water from the

Savannah River production reactors are easily visible in TIR photographs taken from an altitude of 1.2 km. The photographs show that the hot water stream can run more than 2km long and 100m wide and its temperature is at least 5°K above the river ambient.[31] It can therefore be detected by a 40m spatial resolution and 1°K temperature accuracy TIR onboard a satellite (such as MTI satellite). Similarly, the aerial TIR of 50cm resolution is enough to see the warm water stream. Moreover, for the once-through cooling system, another distinguishing characteristic is the cooling water plant and water supply backup portion. Visual or photographic observation could monitor their operational status. In addition, the cooling pond of the reactor should be dried for shut down status, which also can be demonstrated by Open Skies visual examination. Finally, it should be noted that a production reactor has at least several months operating time. This period is long enough for existing satellites with several days revisit time to monitor the hot effluent.

Reprocessing Plants

For overhead imagery monitoring of the operational status of a reprocessing plant, the most important characteristic would be the level of activity at the plant. When operating, there will be many shipments of various forms of nuclear material. Table 4 shows the shipment activities when the PUREX reprocessing plant at Hanford conducted routine operation.[32] If the plant is closed, the activity level should be very low or absent altogether. Transport vehicles, such as trucks, should be big enough to be detected by 1m-resolution images from a satellite. Moreover, if aerial observation is permitted, these activities would be readily seen by visual or photographic images. In addition, airborne effluents (such as Kr-85, I-129) from the stack are typical of an operational reprocessing plant. Thus, one may effectively confirm the status of a plant by using the DIAL system (discussed before). Finally, overhead imagery should be complementary with other forms of measurement, such as the deployment of tamper-proof sensors at key points, environmental monitoring, and on-site inspection.

31. Frank von Hippel, et al., "Stopping the production of fissile materials for weapons," *Scientific American*, 253:3 (1985).

32. G. Thompson, "Verification of a Cut-off in the Production of Fissile Material," *A Handbook of Verification Procedures*, Frank Barnaby, ed. (Basingstoke: Macmillan, 1990).

Table 4. Shipments of nuclear material during operation of the Hanford PUREX plant at 3,000 tonne/year Throughput[33]

Material	Form	Number of Shipments/ year	Transport Mode	Origin	Destination	Total Distance per year[a]
Irradiated fuel	Solid	1125	Railcar	N-Reactor	PUREX	21,600 km
Uranyl nitrate hexahydrate	Liquid	630	Truck tank trailers	PUREX	UO_3 Plant	5,040 km
Plutonium dioxide	Solid powder	360	Diesel truck	PUREX	Z Plant storage	3.331 km
Uranium Oxide	Solid powder	60	Railcar	UO_3	Fernald, OH	230,040 km

Uranium Enrichment Plants

The existing uranium enrichment plants for weapon-grade HEU are mainly gaseous diffusion and gas centrifuge facilities. A gaseous diffusion plant typically has a large size and requires heavy electric power. It usually discharges its heat output through cooling towers. As in the case of reprocessing plants, when a facility is operating, the water-vapor plumes are large enough to be seen by 1m-resolution satellite images. In the context of Open Skies, the operating status will be much easier to detect by airborne VNIR and TIR images. Finally, once a gaseous diffusion enrichment plant is started, its operating time is at least several months. The operating time of a diffusion enrichment plant is long enough for detection by a satellite with several days revisit time.

In the case of a gas centrifuge plant, its smaller size, low power requirement, and lack of a cooling tower mean that overhead imagery could be of less use in detecting heat output. However, the activity level at this site could be observed by overhead imagery. Still, the other monitoring measures should be applied.

It should be noted that the verification measures of a cutoff convention could be different from that of "full-scope" safeguards, because the countries to be verified are different. For IAEA safeguards verification the main target nations are NNWS, and these states have committed their nuclear facilities for the non-production of fissile material for weapons, not only in the future but also in the past. However, for the eight target states, a cutoff treaty could permit their holding of undeclared stockpiles (of past production) and using or

33. Table from G. Thompson, "Verification of a cut-off in the production of fissile material," in *A Handbook of Verification Procedures*, Frank Barnaby, ed. (Macmillan, 1990). All distances are one-way.

processing the earlier-produced fissile material for sensitive military activities (such as the assembly of nuclear weapons). These allowed sensitive production facilities and activities could be co-located with the required verified facilities (such as production reactors and reprocessing plants). Therefore aerial observation or some on-site safeguard verification for some sites might not be permitted, since these monitoring measures could detect sensitive information. Moreover, in order to verify the cutoff convention effectively, some monitoring techniques could be developed and applied — such as the RECOVER (remote continual verification) technique,[34] which is a system for monitoring the performance of certain safeguards sensors (such as the surveillance cameras observing the spent fuel pool of reactors), storing the information, and, upon command, transmitting the information over the international telephone system to some central receiving station. The choice of appropriate means of verification should therefore be based on the criteria of effectiveness, political tolerability, and cost-effectiveness.

Suggestions regarding appropriate organization and management of such a system under the IAEA

From the above discussion, it can be concluded that overhead imagery will play an important role to strengthen the existing safeguards system and the expected cutoff convention, especially to detect clandestine nuclear activities. However, as a multilateral nuclear arms control verification measure, some countries, for reasons of national sovereignty and national security, might not permit aerial observation. In addition, the most difficult task of existing safeguards and environmental monitoring techniques is how to locate undeclared nuclear sites. Aerial observation is less effective for such a case. At the same time, although aerial observation can play some roles to detect clandestine activities at known sites, it can also apply environmental monitoring techniques, which could be the main verification measures in such cases.

A much less intrusive multilateral verification measure is needed and a satellite monitoring system has this advantage. From the view of technology, existing or forthcoming, commercial observation satellites can strengthen the safeguards system. Based on the basic legal regime of the Outer Space Treaty of 1967, satellites for verification are in accordance with the rule that outer space is open to all nations or groups to be used for peaceful purposes.[35] Consequently, observation satellites used as multilateral verification measures are both technically feasible and politically acceptable.

Moreover, these verification satellites should be managed by an international verification organization. The IAEA can use satellite images provided

---

34. F. Prokoski, "Global Monitoring of International Nuclear Safeguards," Proceedings of the Fifth International Conference on Computer Communication, Atlanta, GA, 27-30 October 1980.

35. *Arms Control and Disarmament Agreements: Texts and Histories of the Negations*, 1996 edition.

by some companies or by itself to trigger a special inspection. But this could lead to some disparities. Some nations could buy all the images about their sensitive sites and deny other organizations access to them. The service of some satellite firms (OrbView, for example) will be offered on an exclusive basis. A powerful state could establish itself as the exclusive regional client of multiple suppliers and thus deny others access to images. Additionally, satellite firms might not like to release their best clients' images. The availability of images could also be influenced by a firm's government. In short, from the view of equality, an international satellite verification organization is necessary. In fact, since 1978, France has proposed an International Satellite Monitoring Agency (ISMA). Canada and Sweden have since made similar proposals.[36]

In order to use observation satellites effectively to strengthen IAEA safeguards, the IAEA could manage the satellites specified to monitor nuclear activities, because it is responsible for safeguarding nuclear material production activities. It has a stronger technical capability and has accumulated experience in nuclear safeguards matters. Furthermore, it can support satellite image interpretation by using its data baseline of each member state's nuclear activities. The data from the satellite can be processed, analyzed, and interpreted by the same standard, which would address the problem of erroneous interpretations and faulty judgement.

36. *Ibid.*, ref. 22.

Chapter Thirteen

Northeast Asian Insecurity: New Sources of Friction Between States and Implications for Arms Control and Disarmament

Hong-Kuan Jiang*

Introduction

This chapter will examine the emerging trends of non-traditional security challenges (NTSCs) in the Sea of Japan and the Yellow Sea basin area, and discuss the implications for potential application of arms control and verification technology.

The concept of NTSCs in this study refers to what Robert Bedeski has identified as including "those non-military issues which disturb the national well-being, regional tranquility and international peace."[1] The rationale for this definition is based upon the assumption that non-military threats will play a wider and increasingly important role as sources of international friction at the global level. During the Cold War, the East-West division, with its Soviet-American standoff, either posed structural constraints on, or underrated the impact of, non-military security issues. With the end of the Cold War, the possibility of direct military confrontation between world powers is clearly declining. As Andrew Mack points out, "the declining probability of major interstate war is not accidental, but arises from profound changes in the structure of the international system and in international norms regarding the appropriateness of war as a means of resolving conflict."[2] These profound changes, however, have not resulted in a safe world. Instead, the spread of ethnic conflicts within or across national boundaries, the deteriorating global

---

* This chapter is an edited version of a presentation made to the Fifteenth Annual Ottawa NACD Verification Symposium, "Arms Control and the Rule of Law: A Framework for Peace and Security in Outer Space," 11–14 March 1998, Montebello, Quebec, Canada.

1. Robert Bedeski's working definition in "Integrating Non-Traditional Security Issues: A Preliminary Framework for International Cooperation," in René Unger and F. Ron Cleminson, eds., *Promoting Peace on the Korean Peninsula Through Arms Control: Preventing Regional Conflicts* (Ottawa: Department of Foreign Affairs and International Trade, 1997), 205–16.

2. Andrew Mack, *Proliferation in Northeast Asia*, Occasional Paper no. 28 (The Henry L. Stimson Center, July 1996), 58–59.

environment, the widening gap between haves and have-nots, and increasing illicit activities at the international level, all suggest that the era of mutually assured destruction (MAD) is likely to be replaced by new insecurities arising from a much wider range of sources which are increasingly intrusive to social stability, national governance, and the well-being of humanity.

Northeast Asia, including Japan, South Korea, North Korea, northeast China, Taiwan, Mongolia, and the Russian Far East (RFE), has been an area of primary security concern for decades — perhaps more accurately, for centuries. During the Cold War, Northeast Asian regional security was characterized by two main features: division and confrontation. Confrontation existed not only between the two divided nations — China and Taiwan and South and North Korea, the regional political and ideological rivals — but also between the global powers with heavy stakes in this region. The demise of the Cold War, the collapse of the Soviet Union, and the end of the Sino-Soviet rift with a subsequent normalization between China and Russia, all permitted and invited ground-breaking changes. Within this context, Canada launched the North Pacific Cooperative Security Dialogue (NPCSD) initiative in 1990. The NPCSD, as Dewitt and Job point out, had a clear mandate to increase Canadian security presence in the North Pacific. The NPCSD has been successful in two dimensions. First, it has broadened the concept of security by including non-conventional security issues in which Canada has a role to play. Second, it has established a regional security constituency among security specialists at both the non-governmental and governmental levels (the latter in their private capacity) across the North Pacific with whom Canadians have much to share. "The breadth and level of sophistication of understanding across the range of factors encompassed by the notion of cooperative security" have been increased substantially. "Multilateralism is becoming an important and indispensable part of the emerging security enterprise in Asia Pacific."[3] Many of the NPCSD studies on specific issues have also been instrumental in inspiring new thinking on regional security in Northeast Asia.

Since then, however, changes in Northeast Asia have kept unfolding. The two Koreas gained their UN membership and reached some historical agreements for stability on the peninsula — at least on paper. North Korea began to pursue relations with the US and Japan. It seemed, then, indeed hopeful that the two Koreas might be able to reconcile with each other for a peaceful settlement. Nonetheless, all hopeful signs disappeared quickly after a series of unexpected developments on the peninsula, and tensions began to amount. First came the issue of the North Korean nuclear bomb in 1993 and the standoff between North Korea and the international community; then the sudden death of North Korea's leader — Kim Il-sung — in 1994, and the subsequent political transition in the aftermath. The deeply rooted structural

---

3. David B. Dewitt and Brian L. Job, "Asia Pacific Security Studies in Canada," in Paul M. Evans, ed., *Studying Asia Pacific Security* (Toronto: University of Toronto-York University Joint Center for Asia Pacific Studies, 1994), 86–7.

problems in North Korea aggravated an economic crisis, which was further exacerbated by its worst-ever natural disasters in 1995–96. By 1996, many agreed that it was no longer a question of whether North Korea could survive, but only a matter of when South Korea would absorb the North and under what conditions. Ironically, then came the East Asian financial crisis in late-1997, and South Korea was the one which was hit the most by this region-wide financial turmoil. While South Korea struggles with the consequences of this, especially the difficult and painful structural adjustments required by the International Monetary Fund (IMF) bailout plan, Kim Dae-jun, the new President of South Korea, indicates that Korean unification is not his government's priority. He also wants to assure his North Korean counterparts that South Korea is not going to absorb North Korea. Although Kim Jon-il seems to be in control in North Korea, serious questions still remain about whether the North can survive its deeply-rooted political and economic crises.

On other fronts, the changes are also diverse. While economic regionalization across the Sea of Japan and the Yellow Sea is developing at a fast pace, no substantive progress has been made in building regional security mechanisms. Ambitious multilateral and multibillion-dollar cooperative projects to link energy and resources supplying and consuming countries on the Northeast Asian continent and across the sea have begun to attract serious attention from both the public and the private sectors. However, the lack of political will continues to be the major hurdle impeding comprehensive cooperation. China is rising quickly while Russia is striving for a reversal to its continuous decline. Facing growing regional uncertainties, Japan and the US have renewed their military alliance.

Northeast Asia is indeed entering a period of flux and uncertainty. With regard to the prospects for regional stability, however, analysts disagree and assessments are contradictory. Despite all these controversial developments, it seems increasingly clear that the security dynamics in Northeast Asia are changing. The rising trends of non-military security issues, such as environmental degradation, fishing disputes, transnational migration, and illicit activities, in particular have indicated that it is no longer adequate to focus Northeast Asian security solely on military issues. "National security," defined as political order and social stability, has been challenged by a new range of issues. For security building in Northeast Asia, it is now imperative to address both military and non-military issues, and explore their links as well.

Geopolitics: Changes and Continuities

The post-Cold War geopolitics of Northeast Asia are marked with both profound changes and startling continuities. The end of the Cold War has brought three major changes to Northeast Asian security. First, tensions between former political and ideological rivals have been substantially reduced. South Korea's bilateral relations with its two main political rivals — Russia and China — have been developing on a solid basis of mutual interests. South

Korea and China, in particular, have found common ground not only in establishing closer economic ties but also in addressing many regional security issues such as stability on the Korean peninsula and Japan's future role in Northeast Asia.

What is more fundamental to Northeast Asian security perhaps is the rise of China and its growing strategic partnership with Russia. Bilateral meetings between high-level policymakers in Beijing and Moscow have been conducted on a frequent and regular basis since 1992. The 1992 summit joint statement still cautiously proclaimed that China and Russia would pursue a relationship between "friendly states," which indicates that China and Russia were prepared to remove mutual hostility in bilateral relations. The 1994 summit moved one step further in that China and Russia indicated that they wanted to be "constructive partners," suggesting that existing differences should not affect their interstate relations. The 1996 summit meeting announced that China and Russia would seek a "strategic partnership," which has already prompted speculation.[4] The two countries have almost resolved their century-long territorial disputes in the border area and signed bilateral confidence building measures (CBMs) agreements with the other three central Asian states.

While China and Japan continue to build closer bilateral ties, albeit with periodic tensions, interstate relations between the other two regional powers, Russia and Japan, have also showed indications of potential significant improvement. Russia and Japan are prepared to pursue more constructive and broader bilateral engagement. Thus, except for tensions between the two divided nations, Northeast Asia overall is indeed experiencing relaxed interstate relations for one of the few times in its history. Ideological confrontation between political rivals has been replaced by a growing desire for cooperation in some of the areas which were unthinkable not long ago.[5]

Second, the patron-client relationship which marked the traditional alliance geopolitics in Northeast Asia has shown further signs toward a "more flexible partnership," as identified by Dewitt and Evans in an early study.[6] While it is legitimate to question how much leverage Moscow and Beijing still

---

4. See Liu Jiang, "Another Milestone in Sino-Russian Ties," *Beijing Review's* interview with Igor Rogachev, the Russian Ambassador to China, "Jiang's Russian Visit A Significant Event: Rogachev," and Xia Yishan, "Sino-Russian Partnership Marching into 21st Century," *Beijing Review* 40:18 (5–11 May 1997), 4, 7–8, 9–12. For Russian views, see Yuri Peskov, "Russia and China: Problems and Prospects of Cooperation with CIS Members in Central Asia," *Far Eastern Affairs*, no. 3 (1997), 9–23. Also see Rajan Menon, "The Strategic Convergence between Russia and China," *Survival*, 39:2 (Summer 1997), 101–25.

5. For an analysis on the changing relations among the region's four major powers — China, Japan, Russia, and the US — see Ralph A. Cossa, *The Major Powers in Northeast Asian Security*, McNair Paper 51 (Washington: Institute for National Strategic Studies, National Defense University, August 1996).

6. David Dewitt and Paul M. Evans, *The Changing Dynamics of Asia Pacific Security: A Canadian Perspective*, NPCSD Working Paper no. 3 (January 1993), 2.

have on Pyongyang, it is also justifiable to ask how solid the coalition between Washington, Tokyo, and Seoul still remains, especially given the rising tensions on the broad spectrum of their bilateral relations. On the one hand, these changes mean more room for flexibility indeed, but on the other, they also reflect more ambiguity. As Buzan points out, the patron-client alliance in the region was in essence a structure of "superpower overlay which effectively suppressed indigenous security dynamics ..."[7] What all these changes really mean to Northeast Asian security still requires a "wait-and-see" approach.

Third, the demise of the Cold War and the subsequent developments, as noted above, have also reshaped national security perceptions. In the post-Cold War era, the more threatening challenges to national security, admitted or not as Scalapino has observed, are primarily from internal sources.[8] The uncertainties embedded in the domestic transformation (political, economic, and social) in Russia and China (and also soon in North Korea) have significant security implications not only for national stability in the respective states but also to the region as a whole. For instance, both Russia and China are facing the daunting task of military demobilization and reforming state-owned enterprises. The remaining Russian military forces and defence facilities in the RFE may no longer represent an immediate threat to Russia's Asian neighbours; rather, they become sources of potential troubles for Moscow. "Even the once mighty Soviet/Russian Pacific fleet appears today to threaten only the sailors aboard its ships and the personnel involved in protecting or dismantling its nuclear warheads and propulsion systems."[9]

Despite these important changes, there is a common understanding that Northeast Asian security remains worrisome for two main reasons. One is the continuity of regional fragmentation and division; the other is the increasing uncertainty regarding who will play what role. All multilateral efforts to create regional security mechanisms have not yielded anything of substance.[10] The threat of military confrontation continues between the two Koreas. Political relations between mainland China and Taiwan remain tense, albeit alongside growing economic connections across the Straits. The entrenched Chinese and

---

7. Barry Buzan, *International Security in East Asia in the 21st Century: Options for Japan*, East Asia Project Working Paper Series (University of the Witwatersrand, South Africa, 1996), 8–9.

8. See Robert Scalapino, "Northeast Asia: Prospects for Cooperation," *The Pacific Review*, 5:2 (1992), 101–11.

9. Ralph A. Cossa and Jane Khanna, "East Asia: Economic Interdependence and Regional Security," *International Affairs*, 73:2 (1997), 222.

10. See Paul M. Evans, "Reinventing East Asia," *Harvard International Review* (Spring 1996), 16–19 and 68–9. Also, Peggy F. Meyer, "Post-Cold War Northeast Asia: Regional Trends and Prospects," *The Journal of East Asian Affairs*, 11:1 (Winter/Spring 1997), 65–90; Gerald Segal, "North East Asia," in Robert H. Jackson and Alan James, eds., *States in a Changing World* (Oxford: Clarendon, 1993), 197–218; and Hong-suk Park, "Trilateral Concert in Northeast Asia toward the Korean Peninsula," *Korea and World Affairs* (Spring 1997), 20–43.

Korean resentment against and distrust of Japan continue to exist, if not deepen. The territorial disputes involving many states in the region are likely to carry on into the next century. The traditional notion of national security, which relies upon enhanced military capability, is still prevailing among many Northeast Asian security policy planners. For example, the American decision to deploy theatre missile defence (TMD) systems in the Sea of Japan reflects such concerns;[11] however, this may have increased Chinese suspicions about an American-led ad hoc anti-Chinese coalition in the region. Confidence building at the regional level is as difficult as it ever was. Issues relating to the divided nations continue to produce tensions, trigger periodic confrontation between concerned parties, and threaten the interests of other states in the region as well. Thus, the regional security environment, although improved, remains fragmented.

Geo-economics: Cooperation and Competition

In the post-Cold War era, it is widely conceived that economics has become the defining issue in international relations. Despite the enduring political fragmentation and distrust discussed above, economic concerns are becoming more crucial to national security in Northeast Asia. Over the past several years, economic integration, although uneven geographically and by industrial sector, has progressed at a fast pace at the regional level. It is driven by three major factors. The first factor is the relaxed political relations between states, which tolerate and facilitate transnational economic exchanges. The second is the impact of national economic liberalization in China and Russia, which promotes internationalization at both the national and local level. The third is the growing pressure stemming from intensified global market competition which forces states and localities in the region to opt for cross-border collaboration to maximize national and local comparative advantages.

Three regional trends can be identified to mark the post-Cold War geo-economic transformation in Northeast Asia. First, there is the trend of national economic policy reorientation in nearly all Northeast Asian states to pursue economic interdependence either by making structural adjustments or through adopting more liberal market strategies. In both Moscow and the RFE, Russia is firm in seeking full integration with the dynamic economy in Asia Pacific. After the collapse of the Soviet Union in the early 1990s, Mongolia began actively pursuing economic ties with its Asian neighbours. It supports the enlarged United Nations Development Program (UNDP) regional cooperation project at the Tumen River delta, wishing to build direct transportation links with and gain access to the Sea of Japan. China has been vigorously promoting trans-border connections with all neighboring countries. South Korea adopted the "northern policy" in the late 1980s, and has since been seeking comprehensive relations with China and the USSR/Russia in both political and

11. Stephen A. Cambone, "The United States and Theater Missile Defense in North-east Asia," *Survival*, 39:3 (Autumn 1997), 66–84.

economic realms. South Korean companies have been active in northeast China and the RFE.[12] Japan has been one of the most important players in China's booming economy for two decades, with a strong presence in the old Manchurian area — China's Liaodong peninsula, especially in Dalian — Japan's traditional favorite spot in China. For instance, Liaoning, one of the three provinces in northeast China, has received more Japanese foreign direct investment (FDI) than any other Chinese province over the past fifteen years.[13] During his visit to northeast China in 1997, the first time since the end of the Sino-Japanese War in the 1930s, the Japanese Prime Minister made a commitment of 200 billion yen in government loans to China which, without doubt, will further enhance northeast China's connections with Japan.

Second, intra-regional trade and capital flows have increased drastically. In the late-1980s and the early-1990s, economic transactions were mainly concentrated in southeast coastal China and primarily between the maritime and the continental economies across the South China Sea and East China Sea. Since the early-1990s, there has been a clear northward trend extending to the area surrounding the Yellow Sea and the Sea of Japan. Perhaps the most striking aspect of this trend is the growing economic ties between China and its two Northeast Asian industrialized neighbours — Japan and South Korea. Since the early-1990s, China's bilateral trade with Japan and South Korea has increased sharply. The bilateral trade between China and Japan rose from US$17 billion in 1991 to US$60 billion in 1996, while China's trade with South Korea jumped from US$3 billion to US$20 billion during the same period.[14]

Trade activities on the Northeast Asian continent are also phenomenal. Bilateral trade between Russia and China, albeit primarily in barter, reached its record level of US$8 billion in 1993, nearly double the amount of 1990, and then declined to US$6 billion in 1994. The current level is about US$7 billion annually. Russia keeps a large trade surplus, which provides incentives for barter trade to continue. China's trade with Mongolia and North Korea has gone along a similar path.[15] What may also deserve attention is the increase in Mongolia's economic links with other Northeast Asian states. Mongolia's bilateral trade was once heavily dependent upon the Soviet block countries. Trade with the former USSR alone accounted for about four-fifths of its total foreign trade in the 1980s. It began to change after the Soviet Union collapsed; by 1994, trade with Russia had fallen to about one-fourth of Mongolia's

---

12. For more on South Korean "northern policy," see Kim Hakjoon, *Korea's Relations with Her Neighbors in a Changing World* (Elizabeth: Hollym, 1993), 21–43.

13. See Emily Thornton, "Opportunity Knocks," *Far Eastern Economic Review* (8 December 1994), 56–8, and Shuishi Ono, *Sino-Japanese Economic Relationships*, World Bank Discussion Papers (1992).

14. IMF, *Direction of Trade Statistics Yearbook*, 1997. For an overview on South Korean trade with China, see Hieyeon Keum, "Normalization and After," *Korea and International Affairs* (Winter 1996), 572–89.

15. United Nations ESCAP, *Statistical Yearbook for Asia Pacific*, 1996.

foreign trade. Mongolia's exports to China and Japan grew from 0.7 per cent and 3 per cent in 1990 to 21.6 per cent and 18 per cent respectively in 1994. Mongolia's foreign trade has been significantly reoriented toward Northeast Asian states.[16]

Economic regionalization in East Asia is led by intensified flows of intra-regional investment and offshore relocation of industrial production. In Northeast Asia, intra-regional FDI is also the driving force behind economic growth and plays an important role in nurturing regional production networks. In the 1990s, capital flows across the Yellow Sea and the Sea of Japan have been intensified. Since 1992, China has been one of the largest recipients of Japanese FDI. It is also the largest recipient of Japanese official development aid (ODA) by country.[17] South Korean investment in China grew at an annual rate of 40 per cent between 1993 and 1995. By 1995, China had become the single largest destination of South Korean outflow investment. Quite a large number of giant multinational corporations (MNCs) moved offshore production sites to China's largest city (Shanghai), Shandong and Liaodong peninsulas, and other coastal areas along the Yellow Sea rim. Japanese and South Korean MNCs are taking the lead. What was striking in 1996 is that major South Korean conglomerates, such as Samsung, Daewoo, Hyundai, Sunkyong, and Lucky Goldstar, all made some important moves into China. Nearly half of the South Korean investment in China has been located in Shandong and the old Manchurian area.[18] Shanghai, located at the mouth of the Yangtze River and seen as China's dragonhead, is presently the hottest spot for industrial and service sector investment. What is more important to the changing geo-economics in Northeast Asia perhaps is the emerging regional commercial network, which is centered around Shanghai. It is becoming a regional economic nexus linking other local economic poles in China, Japan, and South Korea. This reflects the central role Shanghai once played as the region's financial and commercial centre in the 19th century.[19] Comparatively, the total amount of inflow FDI in the RFE has thus far been small. Nonetheless, the investment, mainly from South Korea, Japan, and China, has increased noticeably in the extractive, fishery, and service sectors. While Chinese investment is located primarily along the Amur River, South Korean and Japanese investment is concentrated on Sakhalin Island.[20]

---

16. See Department of Foreign Affairs and Trade, Australia, *Pacific Russia: Risks and Rewards* (Parks: Department of Foreign Affairs and Trade, 1996), 41–63.
17. For instance, China received US$1,479 million in Japanese ODA in 1994, which accounts for 18 per cent of the total Japanese ODA for that year.
18. Brian Bridges, "China and the Two Koreas: Searching for a New Balance," in Maurice Brosseau, Kuan Hsin-chi and Y.Y. Kueh, eds., *China Review 1997* (Hong Kong: The Chinese University Press, 1997), 87–106. Also, Keum, "Normalization and After."
19. Furuta Kazuko, "The Shanghai Network Reemerging," *Japan Echo*, 24:1 (1997), 37–41. See also B. Michael Frolic, *China's Second Wave of Development: The Yangtze River Region*, Eastern Asian Policy Papers (Toronto: University of Toronto-York University Joint Centre for Asia Pacific Studies, 1994).

Third, the joint development of energy and resources on the Northeast Asian continent has become the focal point of regional cooperation. Despite the impact of the Asian financial turmoil, which may delay some of the projects, regional cooperation in energy and resources has shown more promising prospects than at any other time in history. Cooperation in the energy and resource sectors is crucial to regional integration in Northeast Asia for two major reasons. The first is that the Northeast Asian continent has tremendous reserves of oil, gas, coal, and minerals, which remain undeveloped. The other reason is the growing market demand in Japan, South Korea, China, and Taiwan — the world's major energy and mineral consumers. During the Cold War, there were a few bilateral projects between Northeast Asian states. The first bilateral agreement in resource development, signed between Japan and former Soviet Union in 1972, was to develop a coking coal mine and coal washing plant at Neryungri, Yakutia. Later, in 1974, South Korea and Japan agreed to jointly develop offshore oil in the waters between the two countries. North Korea and Japan reached an agreement in 1988 to establish a chemical joint venture for extracting and refining rare-earth metals.

The end of the Cold War, the relaxation of inter-state relations and, more importantly, the intensified global market competition for natural resources, all favour multinational cooperation to exploit the vast undeveloped natural resources both on the continent and in the waters. While several major multinational projects have been proposed to develop natural resources on the continent, mainly in the RFE, countries surrounding the Sea of Japan and the Yellow Sea rim are progressively pursuing offshore projects on their own. The Sakhalin Oil and Gas Project, which includes Sakhalin 1, Sakhalin 2, and Sakhalin 3, plans to develop offshore oil and gas east of Sakhalin Island for both domestic consumption and exports to Japan, South Korea, and China. According to some estimates, these fields can produce a total of up to three billion barrels of oil and at least a trillion cubic feet of gas, but more than US$10 billion is needed to make this potential a reality.[21] Russia finally sorted out some long-remaining legal problems regarding rights concerning the Sakhalin projects between the centre and localities in late 1995 and early 1996. The coming summit between Japan and Russia this year will discuss bilateral cooperation, which is expected to give the project a boost.

Several ambitious proposals to build oil and gas pipelines linking resource sites in Siberia, the RFE, eastern Mongolia, and central Asia with the consumer markets in China, South Korea, and Japan have also attracted wide attention. One of the plans proposed by Japanese Tottori Gas Co. is to build a 5,000 kilometre gas pipeline from Irkutsk, through Mongolia, China, the two

---

20. For an overview on foreign investment in the RFE, see Department of Foreign Affairs and Trade, Australia, *Pacific Russia*, 63–76.

21. See James P. Dorian, David Fridley, and Kristin Tressler, "Multilateral Resource Cooperation Among Northeast Asian Countries: Energy and Mineral Joint Venture Projects," *Asian Survey* (Spring 1993), 3–34.

Koreas, and reaching Japan. This pipeline will cost US$10 billion. The Japanese are confident because their estimates of the gas deposits in Irkutsk suggest that there is roughly 850 billion cubic metres of gas, which will be sufficient for Japan's gas consumption for more than ten years. A detailed proposal will be presented to the 1998 Northeast Asia Economic Forum in Yonago. Another proposal is for a 3,000 kilometre oil pipeline between Kazakhstan and Xinjiang in northwest China. This project will cost US$2.3 billion and is designed to transport 25 million tons of oil from Kazakhstan to China. The agreement was signed in May 1997. In addition, there have been serious discussions about building an oil pipeline linking eastern Mongolia's newly found oil field with China's oil giant Daqing in Heilongjiang, which is still the largest oilfield in China producing nearly 40 per cent of the country's oil output.[22]

The UNDP-backed Tumen River Economic Development Area (TREDA) project is also crucial to Northeast Asian cooperation and integration. This project was proposed in 1991 in an attempt to build a subregional transportation hub in the area where China, North Korea, and Russia share their borders. It includes all states in the region.[23] Two institutions have been established: the coordinating committee including Russia, China, and North Korea and the consultative commission with the above three states plus Mongolia and South Korea. Despite many uncertainties, all participant states seemingly remain committed and the project has started its implementation phase. These projects, if successful, will significantly change the strategic landscape of Northeast Asia.

Like the rest of East Asia, economic integration in Northeast Asia is primarily market driven and plays by the "rule of the jungle." Business firms and localities are taking the lead in pushing for bottom-up transnational connections. The long-existing distrust between Northeast Asian states and other "hard" security issues, such as national unification in the case of Korea and China, and territorial disputes, however, are hurdles standing in the way of developing the necessary political will for comprehensive regional cooperation. Within this context, it is indeed important to acknowledge the potential benefits of regional cooperation by pooling together national or local comparative advantages from different states. For instance, each participant will gain from cooperation in combining Russia's natural resources, South Korean and Japanese technology and capital, and Chinese and North Korean abundant labour, with the region's booming economy. Yet, the region's realpolitik also warns that it is essential to recognize that competition and rivalry at the firm, local, and national levels will be likely to affect power redistribution, espe-

22. For an overview of these plans, see Mark J. Valencia, "Energy and Insecurity in Asia," *Survival*, 39:3 (Autumn 1997), 87–9, Alexander's Gas & Oil Connections, "Russia-Japan Pipeline Proposed," http:www.gasandoil.com/goc/company/cns74322.htm (10 February 1998), and *China News Digest* (15 March 1998).

23. Japan is an observer, but has been invited to join the project as a full member.

cially among states, due to a lack of trust and thus makes it harder to achieve cooperation. China, Russia, and Mongolia (and North Korea as well when it opens up) are competing for international capital, technology, and market shares, while Japan, South Korea, and China are striving for secure regional resource supplies. These new relationships between partners and rivals are shaping the new dynamics of regional security in Northeast Asia.

New Dynamics and Regional Insecurities

What has been discussed in the preceding sections about the changing geopolitics and geo-economics are, in essence, several new parallels that have emerged in Northeast Asia. The lessening of political tensions and the improvement of interstate relations are accompanied by growing uncertainties stemming from the political and economic transition within states and shifting power asymmetries at the regional level. National economic liberalization with a growing desire for economic interdependence and cooperation is mixed with intensified regional competition and rivalry. In this context, Northeast Asian insecurity is affected not only by the level of regional integration, but also by the scope of regional competition and rivalry. When addressing broad security issues in Asia Pacific, Funabashi suggests that the focal point should be on the region's growing gap between economic success and the institutional security vacuum. If it cannot be bridged with new checks and balances, the gap will threaten the region's economic success and stability.[24] This can also be applied to Northeast Asia. The new dynamics of political relations between states and the changing power asymmetries will further crystallize national and local concerns regarding who gains and who loses. It is this parallel that gives rise to new regional insecurities and makes NTSCs more acute in Northeast Asia. The narrow conception of national security, which refers only to the absence of external military threat, is no longer sufficient to address security concerns in the region. Given the increasing scope of interaction between states and the growing interdependence in Northeast Asia, insecurities are derived from a much broader spectrum of concerns. For analytical purposes, it is helpful to put them into three broad categories: political insecurity, economic insecurity, and human insecurity.

Political Insecurity

Political insecurity includes concerns in two major areas. One area is the increasing national concern regarding the shifting regional balance of power, which is the direct result of uneven developments between states. China's quick rise has already caused a great deal of concern among its Northeast Asian neighbours. For instance, according to the latest World Bank country study on China, the Chinese economy will expand sevenfold in the next 25 years. This means that China would accomplish in 25 years what took Latin

24. Yoichi Funabashi, "Bridging Asia's Economics-Security Gap," *Survival*, 38:4 (Winter 1996–97), 101–16.

America 65 years and the OECD over 80 years to accomplish.[25] A powerful China can play many different roles in regional security. Japan's move to strengthen its security ties with the United States is also very sensitive to the security of other Asian states. The current crisis in both Koreas does not in any way help reduce the complexity of relations on the Korean peninsula. The future remains as uncertain as it ever was. The gap between China and Taiwan, regardless of growing economic interdependence, is still a major issue of strategic importance between and beyond the two concerned parties. Conventional wisdom would assume that it is in China's strategic interests to see a weak Russia on the other side of its border. Yet, in reality, there are sufficient reasons to support the assumption that a stable Russia could, at the least, assure China that the signed agreement on border demarcation could be implemented. Thus, China may also be concerned about stability in Russia as much as Russia is anxious about the future of China.

The other area relates to the parallel phenomena of growing international integration, on the one hand, and national disintegration on the other. It is important to stress that national disintegration does not necessarily equate to separation. Instead, it refers here more to those structural political challenges that arise from a decentralized economic system and growing local internationalization. Theoretically, national economic liberalization nurtures a more flexible and decentralized system which, as Cox and other critical thinkers forcefully argue, has opened up larger economic spaces for sub-state entities and transnational companies at all levels — global, regional, and local. The change itself will presumably give non-state actors scope "to aspire to greater autonomy or independence, to seek direct relationship to the larger economic spaces..."[26] Therefore, economic regionalization also poses important challenges to national governance. In the case of continental Northeast Asia, open borders and growing local transnational economic ties between the RFE and northeast China represent a test, "with very high stakes" as Rozman argues. Both Russia and China are "struggling with the consequences of transition from Cold War geopolitical reasoning, traditional Communist social organization, and national assumptions about economic interests."[27]

In addition to the above broad structural challenges, there are some practical policy concerns about the dilemma between promoting cross-border economic ties and diffusing rising ethnic tensions. For instance, China has several minorities living along its border with Russia and the central Asian states. When the border was opened for local trade in the early 1990s, it meant

25. The World Bank, *China 2020* (Washington: The World Bank, 1997).

26. Robert Cox, "Structural Issues of Global Governance: Implications for Europe," in Stephen Gill, ed., *Gramsci, Historical Materialism and International Relations* (Cambridge: Cambridge University Press, 1993), 262–3.

27. Gilbert Rozman, "Spontaneity and Direction Along the Russo-Chinese Border," in Stephen Kotkin and David Wolff, eds., *Rediscovering Russia in Asia — Siberia and the Russian Far East* (New York: M.E. Sharpe, 1995), 275.

that it was also open for contacts between ethnic minorities, including separatist movements. The recently increasing separatist activities in Xinjiang, China are closely related to Islamic fundamentalists across the border. Thus, trans-border economic ties at the local level do have implications for national security in such cases.

Economic Insecurity

Economic insecurity stems from a variety of uncertainties inherent in regional cooperation and competition. As Zoellick points out, "successful economies require competition, and competition produces winners and losers."[28] The most worrisome area in Northeast Asia is the competition for energy resources. With an expanding economy, energy supplies will become more crucial in most of the states. Japan, China, South Korea, and Taiwan had already accounted for more than 65 per cent of the total oil consumption in the Asia Pacific region by 1995. Except for China, all of them have relied almost entirely on imports to meet domestic needs. China, however, has become a net oil importer since 1993, but its demand is expected to expand drastically in the next decade or so. According to Standard & Poor's energy forecast, by 2020 China's oil consumption will be three times the amount of 1995 and 50 per cent of this consumption will rely on foreign supplies.[29] In 1995, China spent US$5 billion for 35 million tons of oil and gas imports. By 2020, the amount will be about US$28 billion at today's prices, which should not be a financial problem at all, given that the consumption forecast is based on a 6.7 per cent annual growth in GDP.[30] Japan, South Korea, and China have all taken an active role in the joint development of oil and gas in the RFE. What remains a concern, however, is how to secure long-term and stable overseas supplies. The safety of energy supplies will be one of the key security concerns in Northeast Asia. It is a practical security matter concerning not only how to ensure that the projected multinational oil and gas pipelines will be free from potential disruption, but also how to secure the transportation of sea lanes which are presently vulnerable, especially given the maritime disputes existing in the region and the absence of conflict management mechanisms.[31]

Offshore oil exploration in the Sea of Japan and the Yellow Sea area, especially in disputed waters, is increasingly sensitive. During the Cold War, the Sea of Japan and the Yellow Sea between the continental and the maritime

---

28. Robert B. Zoellick, "Economics and Security in the Changing Asia-Pacific," *Survival*, 39:4 (Winter 1997–8), 29–51.

29. Standard & Poor's DRI Energy Group, "Asia/Pacific Outlook," http://www.dri.mcgraw-hill.com/energy/ap0297.htm.

30. The World Bank, *China 2020*, 50.

31. For a detailed study on energy and insecurity in Northeast Asia, see Valencia, "Energy and Insecurity in Asia," and also, Keun Wook Paik, *Gas and Oil in North-east Asia: Policies, Projects and Prospects* (London: Royal Institute of International Affairs, 1995).

states, were a "virtual no-man's land." There were very few offshore exploration activities in the region. Now every state along the coast is interested in and concerned about offshore oil explorations. After the adoption of the 200 nautical mile exclusive economic zones (EEZs) at the Third UN Convention on the Law of the Sea in 1982, a number of Northeast Asian states have subsequently claimed their EEZs in the region. As a result, some of the waters have actually been claimed by several states. Given the unsettled territorial disputes over several islands between Russia and Japan in the Sea of Japan, two islands between South Korea and Japan, and the Diaoyu or Senkakus island between Japan, China, and Taiwan, the potential for maritime confrontation is obvious. Disputes are not only about sovereignty, but also about resources under the water. As Valencia points out, maritime frontiers could become a "dangerous no-man's land" and island and maritime disputes could reemerge as "hot spots."[32]

Human Insecurity

Human insecurity is another new dimension of regional security which addresses concerns at the individual level. The increasing interactions between states are not merely exchanges of goods and flows of capital between different geographical locations, but also communications between people. The opening of national borders brings to citizens opportunities for prosperity on the one hand, but, on the other hand, they also gain more chances to be exposed to a variety of new threats which are directly or indirectly from sources across the border. Such threats often include pollution, diseases, migration-related issues, and international crimes. For instance, foreign investment in the logging sector in the RFE has been widely seen as being beneficial to Russia. Yet, the large-scale investment in this sector without concrete ecological protection measures will result in over-logging. One of the consequences is deforestation and, as well, the lives of local Russians will be affected directly. Hyundai, a South Korean industrial giant, invested US$16 million in 1993 to establish a logging operation in northeast Primorsky and planned to export its products to Japan and South Korea. With an increasing awareness of the effect of deforestation, local residents began to press Russian governments at all levels and protested against Hyundai's operation in that area. Under public pressure, the Russian federal and local governments took protective measures by increasing taxes on raw material exports. What this case has revealed is not merely the friction between Russians and foreign companies which, to some extent, is also new, but more importantly, the increasing importance of the dimension of human security in Northeast Asia.

While bilateral relations between Moscow and Beijing have indeed improved substantially over the past several years, ironically, ethnic tensions between Russians and Chinese have escalated to a level unheard of for many years. For instance, according to the Director of the Institute of History in

32. Valencia, "Energy and Insecurity," 94.

Vladivostok who conducted two local surveys in 1992 and 1993 respectively, local Russians in the Far East saw the Chinese as irresponsible, dishonest, and rude. Only 4 per cent of local Russian officials and 2 per cent of the Russian public surveyed preferred to work with the Chinese. Chinese products in the Russian surveys were rated as the least welcome foreign goods, with "a virtual zero" per cent rating. The Chinese appeared "to be pursuing nothing less than a shift in surplus population from northeast China to the RFE to form a Chinese diaspora with security property rights ..." The flood of Chinese labourers to the RFE was seen as a new trend leading to another "yellow peril."[33]

Although the reasons behind the rising ethnic tension are complicated, the Russian perception of human insecurity has played an important role. The opening of the border in the late 1980s and the early 1990s and the expanding economic ties between the two sides have generated three major concerns on the Russian side. The first is the increasing pressure of Chinese migration. With such a massive population living next door, the opening of the border presents an immediate threat of potential large-scale migration. Although Moscow had an agreement with Beijing in the early 1990s to use the Chinese "surplus" labour to help develop the RFE, the sensitivity drastically increased when a large percentage of the Russian population became unemployed during the market transition. The second major concern is about border trade with China. Many Russians are not satisfied with China's cheap but inferior consumer goods. Some are frustrated and believe that border trade is not fair. Russia sells natural resources to China in exchange for poor-quality Chinese products. The third major concern is over increasing criminal activities and the deteriorating environment on the Russian side, both of which are believed to be related to the large number of illegal Chinese migrants and Chinese trade practices. As Bedeski points out, "[I]ndividuals represent the foundation of human society, and any policy which omits them is not only incomplete but dooms social discourse to failure."[34] Thus, human insecurity has become an important component of the broadened concept of security in post-Cold War Northeast Asia where even traditional security issues are still paramount.

New Sources of Maritime International Friction

Having discussed the changing dynamics of regional security and outlined the emerging multilevel insecurities in Northeast Asia, this section will identify sources of maritime international friction in the Sea of Japan and the Yellow Sea basin area. Maritime NTSCs are not new in this region. In terms of the

---

33. Viktor Larin, "Yellow Peril Again?" in Stephen Kotkin and David Wolff, eds., *Rediscovering Russia in Asia — Siberia and the Russian Far East* (New York: M.E. Sharpe, 1995), 295. For recent surveys see Russian Far East Update, *The Russian Far East: A Business Reference Guide*, 2nd Edition (Seattle: Russian Far East Update, 1996), 157.

34. Bedeski, "Integrating Non-Traditional Security Issues," 212.

impact on inter-state security relations, however, they are gaining importance and becoming new elements affecting changes in the regional order or disorder. Maritime NTSCs are likely to arise from the following four major issue areas.

Fishing Rights

Disputes over fishing rights are an old issue in the Sea of Japan and the Yellow Sea rim. The Sea of Japan and its adjacent waters have long been known as one of the areas in the world with rich fishing resources, and thus one of the most productive areas for the fishing industry. According to Hayes and Zarsky, the North Pacific region accounted for 32 per cent of the world catch in 1984, out of which almost 90 per cent was caught in the northwest Pacific. Countries, especially localities surrounding the Sea of Japan have developed the fishing industry into a major economic sector. The fishing industry in South Korea, North Korea, and Japan is heavily dependent on fish stocks in this portion of the water, which provides about 90 per cent of their catch.[35] Although Russia's catch from this area is about 30 per cent of its national total, it is still one of the most important economic sectors in the RFE and produces exports, attracts foreign investment, and provides jobs. Dalmoreproduct, a former state-owned fishing company and now a holding firm in Vladivostok, for instance, has 14,000 employees. Seafood items keep a share of about 40 per cent of the RFE's total exports to Asian countries.[36] Since the early 1990s, the fishing industry in the RFE has been contracting. The 1994 catch was only about 2.2 million metric tons, less than 50 per cent of the 5 million in 1988. However, it remains a major local economic sector.[37]

The prominent position of the fishing industry in this area, the migratory nature of fish resources, the increasing catch activities, and the ambiguity in fishing rights jurisdiction are all-important factors conducive to growing concerns and periodic tensions between neighbouring countries. For instance, high seas Pollack fishing disputes between Russia, South Korea, China, and Poland, the main catchers in the central waters, trigger on-and-off tensions between these states. In 1993, Russia detained a Chinese trawler in this area while China retaliated by boarding Russian fishing vessels. Both sides got very anxious and relations were tense. Such incidents also occurred between Russia and South Korea, Japan, and Poland. Thus, the interaction between states on fisheries in the Sea of Japan and the Yellow Sea is always an important element of formal diplomacy.[38] Regulations are mostly based upon

---

35. Peter Hayes and Lyuba Zarsky, "Environmental Issues and Regimes in Northeast Asia," *International Environment Affairs*, 6:4 (1994), 289.

36. Department of Foreign Affairs and Trade, Australia, *Pacific Russia*, 42.

37. Russian Far East Update, *The Russian Far East*, 132–6.

38. See Douglas M. Johnson and Mark J. Valencia, "Fisheries," in Mark J. Valencia, ed., *The Russian Far East in Transition* (Boulder: Westview Press, 1995), 147-82, and also

bilateral agreements which need to be renegotiated and renewed after a certain period of time.

At present, there are several concerns over fisheries in the Sea of Japan and the Yellow Sea basin area which are likely to generate new regional frictions, if they are not handled properly and promptly. One concern is over the declining stocks due to over-fishing, which threatens both the national and local fishing industry. Disputes are often related to fishing grounds, the number and size of vessels to be licensed, the catch to be allowed including catch quotas for major species, and fishing on the high seas. Complaints cluster over catching standards, violations of the agreed quotas, different assessments of fishing stocks, and national protection measures on certain species. The disagreement between Japan and Russia displayed in the negotiation on the renewal of their bilateral fishing accord is a recent typical case.[39]

Another source of potential friction is the controversy resulting from the legal concept of the EEZ which allows coastal states to claim a 200-nautical mile extension of the baseline of the territorial sea. There have been some collaborative efforts to establish effective regulatory regimes in the region. One is the Anadromous Species Commission which includes Russia, Japan, and the US. The other is the International Council for the Exploration of the Sea in the North Pacific (PICES) which was established in 1994 and includes Russia, China, Japan, and the US. While it is not quite clear yet whether these regimes will be more effective, their limits are clear from the beginning. More importantly, several active players in the region are not included.[40] While China and Japan have reached a bilateral agreement regarding fishing rights in the Sea of Japan and East China Sea, South Korea and Japan disagreed on waters surrounding the two disputed islands between them. When Japan unilaterally abolished the 1967 bilateral accord in 1998, tension began mounting. Several incidents almost triggered a diplomatic crisis.

When South Korean fishing vessels appeared on Japan's newly claimed contingent waters in January 1998, Japan sent its coastal guards, captured all South Korean fishing vessels defying their orders, and arrested all the crews on board. South Koreans took to the streets and surrounded the Japanese Embassy in Seoul with angry anti-Japanese protests. The South Korean government had to send in armed police.[41] This happened at a time when South

---

Douglas M. Johnston, *Fishery Politics and Diplomacy in the North Pacific: New Sources of Instability*, NPCSD Working Paper no. 7 (Toronto: York Centre for International and Strategic Studies, 1992).

39. For background information, see Tadashi Yamamoto and Hajime Amanishi, "Use of Shared Stocks in the Northwest Pacific Ocean with Particular Reference to Japan and the USSR," in J. Marsh, ed., *Resources and Environment in Asia's Marine Sector* (London: Taylor and Francis, 1992).

40. For a detailed discussion on the impact of the EEZ, see Johnson and Valencia, "Fisheries."

41. "Tokyo Asks Seoul to Drop Bilateral Fishing Pact," *The Japan Times*, International Edition (26 January–1 February 1998), 2. "Seoul Criticizes Japan's Seizure of ROK

Korea was in a serious financial crisis, and Japan was expected to be one of the major participants of the IMF-backed bailout plan. Given resurgent and strong Korean nationalism during the present financial turmoil, it could become an explosive issue. While the South Koreans accuse the Japanese of taking advantage of South Korea's vulnerable position, the Japanese blame the South Koreans for violating their newly claimed 200 nautical mile EEZ. Both the South Korean President and the Japanese Prime Minister have called on the other side for a "stern" and "future-oriented" response, and do not wish to see their bilateral relations damaged by irrational and undesirable actions from the other side. While it is hard to imagine that fishing disputes can lead the two states to a large-scale military confrontation in the Sea of Japan, it is not surprising, however, that the tension can escalate to an unexpected level that may hurt bilateral relations between Seoul and Tokyo. Both sides have been cautious and kept the door open for further negotiation.

Maritime Pollution

As Scalapino has argued, the greatest security threat in the next century will arise from environmental degradation.[42] In the Sea of Japan and the Yellow Sea area, there are three major concerns: air pollution, waste dumping, and oil spills.

Air pollution refers mainly to acid rain, which is one of the major pollution sources in the Sea of Japan and the Yellow Sea rim. Acid rain mostly results from high levels of sulphur emission, which pollutes lakes and rivers, and destroys crops, forest, and fish stocks. The World Bank country study on China notes that acid rain "threatens to damage 10% of the land area and may have already reduced crop and forestry productivity by 3%" in the high-sulphur coal regions of southern and southwest China.[43] What is more critical is that acid rain is mobile and has a trans-boundary causal effect. An increasing number of studies by South Korea and Japan have found that the sources for the high-level emissions over Japan and South Korea are in China, because China is the major source for high-level sulphur emission through its coal burning. Northeast China usually has more serious particulate pollution because of the large amount of coal consumption. Recent joint research

---

Trawler," and "Leaders Seek to Bring Calm to Fisheries Stalemate," *China Daily* (22 January 1998), 12 and (27 January 1998), 12, respectively.

42. The Asian Society, "Environmental Issues," *Prospects for International Cooperation in Northeast Asian Multilateral Dialogue* (New York: The Asia Society, 1995), 9. For comprehensive studies on security implications of environmental issues, see Thomas Homer-Dixon, "Environmental Scarcities and Violent Conflict: Evidences from Cases," *International Security*, 19:1 (1994), 5–40; Shakut Hassan, "Environmental Issues and Security in South Asia," *Adephi Papers*, no. 262 (London: IISS, 1991), and also Ashok Swain, "Displacing the Conflict: Environmental Destruction in Bangladesh and Ethnic Conflict in India," *Journal of Peace Research*, 33:2 (1996), 189–204.

43. The World Bank, *China 2020* (Washington: The World Bank, 1997), 2.

conducted by several local Japanese environmental laboratories, through computer simulations of acid rain-causing pollutants in the region, has found that the acidity of rain or snow over the Sea of Japan becomes higher in winter when the seasonal wind from China to Japan strengthens.[44]

This study further reinforces the existing perception that China is the region's major air polluting source and thus should bear more responsibility for cleanup. China argues, however, that it is not the only polluter, but rather the latest in the region. Other states at an advanced stage of development thus should not mitigate their accountability for the declining and threatening regional environment. No matter who is responsible for the air pollution, the deteriorating environment threatens everyone in the region. The World Bank's China report has released some alarming figures about the environmental cost of development in China: 178,000 people in major cities suffer premature deaths each year because of pollution; indoor air pollution from coal and biomass from cooking and heating causes 110,000 premature deaths each year in rural areas; water pollution has contaminated 52 of the 135 monitored urban river sections; acid rain threatens to damage 10 per cent of the land area; and air and water pollution's damage to human health has been estimated at US$54 billion a year, which equals 8 per cent of China's GDP in 1995.[45] The report has offered China two options: business as usual or a new alternative which will have completely different results. Whether China can follow the World Bank prescription and whether the World Bank prescription can work will depend upon China's domestic development as much as regional and international joint actions. Yet, these findings will perhaps enhance regional collaborative efforts and strengthen the few existing regional environmental regimes.

The second issue of maritime pollution is the unregulated *waste dumping* activities in the Sea of Japan and the Yellow Sea area, including sewage discharge and radioactive waste dumping. It is a common practice in each coastal state to dump wastes into the water. However, the present waste dumping activities mainly occur along the west coast of the Chinese and Russian side where there are very few waste treatment facilities. Dumping waste into the Yellow Sea has been a routine activity of local Chinese industrial firms for years. According to some recent Chinese estimates, 50 million cubic metres of solid waste are discharged into the sea from the Chinese side annually. If a ship carries an average of 700 cubic meters, there would be approximately 70,000 ships to discharge waste into the sea from the Chinese side each year. Although there has been no study thus far to verify the concrete environmental cost of this amount of solid waste discharged into the water, there is no doubt that the effect of such practices with other pollution sources will significantly change the ecological system in the Yellow Sea basin area.

---

44. See "Japanese Researchers Simulate Acid Rain in East Asia," *Global NewsBank* (Kyoto World Service) (19 August 1997).

45. The World Bank, *China 2020*, 2.

It has caused serious concerns in South Korea. As well, the Chinese State Oceanographic Administration, the national agency responsible for maritime regulations, plans to impose a new system this year, expecting to bring the current chaotic dumping activities under control.[46]

The UNDP Tumen project has conducted some comprehensive studies on the ecosystem in the TREDA. It has found that the marine ecosystem near the greater Vladivostok area is affected by industrial wastes, hydrocarbons, sewage, and sedimentation. Except for two plants in the suburbs of Vladivostok, there are no sewage treatment facilities in the whole TREDA. This means that sewage is discharged into the sea without any treatment.[47] In contrast to the Vladivostok area, the southeast coast part of the RFE is not densely populated and has a well-established and well-designed system of marine and territorial reserves. Despite the impact of the lack of funding under Russia's current circumstances, the Posyet Bay, with high levels of biodiversity, is still the most productive area along the coast of the Sea of Japan. Nonetheless, it is now becoming sensitive to a number of pollutant sources on both the Chinese and the North Korean side because of the increasing economic activities near the border area. The north-flowing Korean current in the Posyet Bay and the Tumen River are the two main channels which move pollution into this sensitive area. A 1997 report by the Vladivostok Institute of Sea Biology states that the water coming from the Tumen River to the Sea of Japan through the Posyet Bay has been polluted by phenol, metals, and other hazardous substances. These pollutants are mostly from China where the local population is dense and few enterprises have purification facilities. Thus, the report suggests that Russia should insist on ecological security as a priority in implementing the UNDP Tumen project.[48]

Another closely related waste dumping issue is the former Soviet Navy's radioactive discharge into the Sea of Japan. This has been a major Japanese concern for some time. In 1993, Russia finally admitted that between 1978 and 1992, the former Soviet and current Russian Navy had dumped 18 decommissioned nuclear reactors and 13,150 containers of radioactive waste, mostly into the Sea of Japan.[49] Neither the Japanese nor the South Korean

46. "Control of Dumping at Sea to Be Tightened," *China Daily* (16 December 1997).
47. The Tumen River Area Development Program Environment Sub-Group, the UNDP, "Report G: Preliminary Environmental Study," *Collected Papers* (1994).
48. "Tumen River Brings Polluted Water to Vladivostok Coast," *ITAR-Tass* (25 March 1997). For a detailed study of the political economy of the UNDP Tumen regional development project, see Hong-kuan Jiang, "Borders, Boundaries, and Connections: The Political Economy of Northeast Asian Trans-border Cooperation" (unpublished Ph.D dissertation, York University, Toronto, 1997).
49. Cited in Hayes and Zarsky in "Environmental Issues and Regimes in Northeast Asia," from: the Administration of the President of the Russian Federation, *Facts and Problems Related to the Dumping of Radioactive Waste in the Seas Surrounding the Territory of the Russian Federation* (24 October 1992), translated by Green Peace Russia (April 1993).

reaction of outrage, nor Russia's own admission, did much to stop this practice. In late-1996, it was reported that one Russian naval vessel dumped nearly 1,000 tons of low-level waste. What is more critical is the growing regional concern about Russia's capability to handle the remaining nuclear reactors and fuel from decommissioned nuclear-powered submarines, given its deteriorating economy. Ironically, Russia is not the only one dumping radioactive waste into the Sea of Japan; one Japanese source revealed that Japan dumps much more radioactive waste into the water than Russia.[50]

In 1997, North Korea and Taiwan signed an agreement to allow Taiwan to export its nuclear waste to North Korea. This deal has raised serious concerns in both China and South Korea. While South Korea is mainly concerned about the capability of North Korea in handling nuclear waste, China has an additional concern about the political impact of the agreement on Taiwan-North Korean relations. China argues that only a sovereign state (that is, Taiwan is not a sovereign state) has the legitimate right to sign international agreements on nuclear related-matters.[51]

The third issue is *oil pollution* which includes routine discharges of oil tank washing and accidental oil spills. According to some estimates, about 0.35 to 0.5 per cent of the crude oil in the tankers will settle at the bottom after a long shipping journey. All tanks have to be washed, and so the wash-out of oil is usually discharged into the sea. Such concerns are mainly in Southeast Asia at present where the sea lane traffic is heavy with oil tankers.[52] Yet, it is expected to become an issue soon, given the growing oil shipping sea lane traffic in the Sea of Japan and the Yellow Sea. The more serious oil pollution issue in this area is from accidental oil spills. In recent years, such incidents have increased while more oil tankers are traveling between different seaports in this area. For instance, the number of oil spill accidents reported from South Korean sources increased from 152 in 1987 to 240 in 1991, and the amount of oil spilled into the water almost tripled.[53] The damage in some cases can be catastrophic. *Nakhodka*, a Russian oil tanker carrying crude oil from Shanghai to Kamchatka in the RFE, broke apart near a small fishing town on the Japanese west coast in January 1997 and leaked about 5,580 tons of oil. This incident, although still uncertain about its long-term damage, destroyed the town's only economic source — green seaweed catch. The cleanup was estimated to take more than a year.[54] In July 1997, a Panamanian-registered

50. Valencia, "Energy and Insecurity in Asia," 102.

51. Xiang Jiagu, "Situation on the Korean Peninsula," in René Unger and F. Ron Cleminson, eds., *Promoting Peace on the Korean Peninsula Through Arms Control: Preventing Regional Conflicts* (Ottawa: Department of Foreign Affairs and International Trade, 1997), 28.

52. For more technical information about different ways to measure oil pollution, see Hayes and Zarsky, "Environmental Issues and Regimes in Northeast Asia," 287, and for a regional overview, see Valencia, "Energy and Insecurity in Asia," 100–2.

53. Hayes and Zarsky, "Environmental Issues and Regimes in Northeast Asia," 288.

54. Eugene Moosa, "Troubled Waters Remain in Japan's Worst Oil Spill," *Global News-*

supertanker carrying 257,000 tons of crude oil from the Middle East headed towards Kawasaki, Japan, and ran aground in Tokyo Bay. With about 12,000 tons of crude oil leaking into the water, it became one of the biggest oil spills near Japan. A special task force had to be established, including Japan's Self-Defense Forces.[55] The increasing frequency of oil spilling in these waters is becoming a serious issue that each state along the coast has to deal with.

Cross-Border Migration

Although cross-border migration does not present an immediate threat in the region, the pressure of potential trans-boundary movements of people is always a major concern in several states. The substantial differences existing in resource endowments and levels of development between countries surrounding the Sea of Japan and the Yellow Sea are, and will continue to be, the major driving forces behind potential massive cross-border migration.[56]

At present, the states at the low-income level can maintain relatively strict control over movements of people beyond their national boundaries, while higher-income level states can keep inflow migrants at a manageable level, mainly through tightening immigration regulations. Most of the contemporary concerns focus on illegal migrants and the potential flux of immigrants or refugees from countries where the government may lose control or the system may collapse. Speculations in this area have focused primarily on the possible outflows of massive migrations from North Korea if the regime collapses and from China if its risky reforms fail. In addition, Japan has expressed serious concerns about increasing illegal migration from neighboring countries.[57] Russia has taken tough measures to limit Chinese migrants, both legal and illegal, to a level as low as possible, despite the pain it may cause to the economy.[58]

Illicit Activities

International illicit criminal activities include cross-border smuggling, piracy on the high seas, transnational drug trafficking, and international terrorism. These activities challenge the capability of the state and the well-being of its

---

Bank (8 September 1997).

55. "Tanker Taken in Tow after Big Crude Oil Spill," *Kyodo News International* (2 July 1997).

56. For a comprehensive study on migration in the region, see Won Bae Kim, "Migration," in Mark J. Valencia, ed., *The Russian far East in Transition* (Boulder: Westview Press, 1995), 65–86, and "Sino-Russian Relations and Chinese Workers in the Russian Far East," *Asian Survey*, 34:12 (December 1994).

57. See Steven A. Spencer, "Illegal Migrant Laborers in Japan," *International Migration Review*, 26 (1992), 754–86, and Haruo Shimada, "The Employment of Foreign Labor in Japan," *Annals of the American Academy of Political and Social Science*, vol. 513 (1991), 117–38.

58. For a more objective analysis, see Vladimir Yakovlevich Portyakov, "Are the Chinese Coming?" *International Affairs*, 42:1 (January/February 1996), 132–140.

citizens. They also often generate interstate frictions. In general, illicit activities flourish under three conditions: when the state is weak; when the state is involved directly or indirectly; and, when political control is lessened and transnational interaction increases. The difficulty to identify the circumstances of particular cases is often the source of friction.

After the end of the Cold War, it has been noticed that the pattern of illicit activities in Northeast Asia has changed. While terrorist activities at the regional level decline, smuggling, piracy, international organized crime, drug trafficking, and other activities are on the rise, especially around the Sea of Japan and the Yellow Sea basin area. For instance, smuggling is very active on the Northeast Asian continent and on the high seas. There have been reports of increasing piracy on the high seas.[59] At present, regional concerns are primarily in the following two areas:

Drug trafficking. As Oh's study has found, Northeast Asia has become "a conduit for consignments" of heroin and marijuana destined for North America, Europe, and Australia. Southwest China near the golden triangle has been identified as one of the major sources.[60] China, Japan, and South Korea all have reported a substantial increase in drug cases over the past several years. Japanese official sources confirmed that the amount of drugs seized and the number of people arrested in 1996 increased, respectively, sevenfold and 13 per cent from the previous year.[61] Recently, Russian sources also confirmed that North Korea is another source of drug trafficking in Northeast Asia. North Korean diplomats are directly involved. Over the past few years, more than 30 drug cases involving North Koreans, including diplomats, have been reported. North Korean defectors have also revealed that drug production in North Korea is nearly a state-run business.[62]

Organized crime. Asia, especially East Asia, and Russia were once famous for organized crime. During the Cold War and under the totalitarian government, organized crime was largely suppressed. When economic liberalization began and political control was lessened, organized crime started to emerge and grow in East Asia and Russia. According to an American Congress report, there are four major international organized crime groups in the region. They are the Hong Kong-based triads, the Taiwan-based triads, the Japanese

59. See Valencia, "Energy and Insecurity in Asia," 99.

60. For a detailed study on international illicit activities in Northeast Asia, see Robert E. Bedeski and Kwan-Chi Oh, "The Application of Arms Control Monitoring and Verification Technology and Methodology to Non-Traditional Security," in Robert E. Bedeski, ed., *Confidence Building in the North Pacific: New Approaches to the Korean Peninsula in the Multilateral Context* (Ottawa: Department of Foreign Affairs and International Trade, 1996), 219–33.

61. "96 Haul of Stimulant Drugs Shows 7-Fold Rise: Report," *Kyodo News International* (20 December 1997).

62. *Northeast Asia Peace and Security Network Daily Report* (The Nautilus Institute, 3 March 1998).

yakuza, and the Russian mafia.[63] These organized crime groups are also involved in many international illicit activities, such as drug trafficking, smuggling, and other criminal activities.

Potential Applications of Arms Control and Verification Technology

The preceding sections have discussed the changing geopolitics and geo-economics in post-Cold War Northeast Asia and highlighted how the interaction between changes and continuities has nurtured new security dynamics. It has been argued that the NTSCs identified in the Sea of Japan and the Yellow Sea basin area are not only challenging the national security of the state, but also posing threats to the well-being of its citizens. Hence, the management of NTSCs is becoming a critical issue to vital security interests in the region as a whole. Despite the fact that unilateral policy measures and bilateral collaboration are necessary and indispensable to deal with NTSCs, regional cooperation is widely seen as the most crucial venue to build more effective management mechanisms. This is because many of the NTSCs have trans-boundary effects and threaten all states in the region. Thus, the problems of management often transcend the capacity of any national government, no matter how concerned, committed, or active any particular state is in national management. In this regard, confidence building at the regional level is extremely important to the success of multinational negotiation and cooperation. International confidence building in general can be seen as a process which involves a wide range of collaborative interaction between and among different players. It also relies on a set of agreed norms and verification technologies to ensure the compliance of the signatories. Thus, it is important to examine the potential applications of arms control and verification technologies.

From a security perspective, Hayes and Zarsky have suggested five critical issues that arise from regional environmental regime building in Northeast Asia which can offer insights into this question. The five critical issues are: capacity building; spillover effects between environmental issues; monitoring and enforcement; financing; and, the role of NGOs.[64] Here, I would like to stress in particular the issues of capacity building and monitoring and enforcement, both of which are the key components of successful arms control and disarmament regimes.

Confidence building and capacity building are intertwined. In general, when confidence is gained between different players, their capacity to accommodate differences is also enhanced. Capacity building includes a variety of activities and programmes at both the national and the international level

63. US Congress, Senate Committee on Government Affairs, *Asian Organized Crime: The New International Criminal*, Hearing before the Permanent Subcommittee on Investigations (Washington: US Government Printing Office, 1992).

64. Hayes and Zarsky, "Environmental Issues and Regimes in Northeast Asia," 306–11.

which aim at enhancing the national oversight capability of identifying and addressing NTSCs, and developing national and international institutions to be able to bring together "the different stakeholders." Given the broad scope of NTSCs and the diversity of concerns and interests in Northeast Asia, capacity building is likely to be more critical to regional cooperation and institutionalization. Over the past four decades, international arms control and disarmament have established relatively well-structured procedures regarding how to approach "hard" security issues and developed a set of norms concerning the enforcement, compliance, and verification of agreed treaties or covenants through inspection and transparency. The arms control and disarmament community, although primarily in the West, has also gained sophisticated experience of international regime and coalition building. Thus, it is plausible and valuable to explore further the applicability of confidence building in arms control and disarmament to specific cases of NTSCs in the region.

Regional cooperation on NTSCs also requires all parties to accept intrusive monitoring, transparency, and verification of national codes, action plans, manufacturing activities, and trade practices to ensure the enforcement of compliance to agreed regional standards. These principles have been proven to be an indispensable part of a successful regime in arms control and disarmament. Verification, for instance, is in essence a subsystem to support and guarantee confidence building in arms control and disarmament. A verification regime often includes "measures, procedures and methods for acquiring the information necessary to assure compliance, deter non-compliance and/or resolve ambiguous events."[65] In this regard, arms control and disarmament can also offer valuable experience in helping establish operational mechanisms to support regional non-military security regimes.

Furthermore, regional cooperation on NTSCs involves high-level technical capability, not only for monitoring the compliance of the agreed parties as discussed above, but more importantly for detecting threat sources and assessing damages. It is often the case that only a few states have the capability to identify the sources of NTSCs, or to appraise the degree of threat or the potential damage. In the field of arms control and disarmament, for instance, there is the International Satellite Monitoring Agency with an early-warning system. For an effective identification and appraisal mechanism of NTSCs, perhaps it is also essential to establish a regional or global agency with such capability to detect, monitor, and provide early-warnings. Satellites have already been used in a few countries to detect natural disasters and follow climatic changes and even the migration patterns of animals. Thus, it should not be a problem of feasibility but rather an issue of technology diffusion, information access, and political will to cooperate.

---

65. Patricia Bliss McFate, et al., *The Converging Roles of Arms Control Verification, Confidence-Building Measures, and Peace Operations: Opportunities for Harmonization and Synergies* (Ottawa: Department of Foreign Affairs and International Trade, 1994).

This chapter represents a preliminary attempt to study Northeast Asian regional security by developing a framework which is able to reflect the changing security dynamics in the region, and to explore the potential application of arms control and disarmament to the management of NTSCs. Further research work needs to be done in at least two major areas. First, it is unclear what the spillover effect, if any, is between "hard" security issues and NTSCs. Given the unique security context of Northeast Asia, it is important to investigate the potential links between regional collaborative efforts on NTSCs and regional security institution building. For instance, there have emerged four regional environmental regimes: Northwest Pacific Action Plan; Intergovernmental Oceanographic Commission; Northeast Asian Environment Program; and, the UNDP Subregional Technical Cooperation and Development.[66] There are also a number of regional regimes on fisheries. What will be the role of these functional regimes in regional security building? What will be the impact on arms control and disarmament in Northeast Asia? Second, the distinct features of different NTSCs require detailed case studies to explore the application of arms control and disarmament technologies to specific cases. Although confidence building measures are applicable in general to all regional cooperative regimes, it seems that different issues, such as fishing disputes and maritime pollution, may underline certain specific technologies and skills to build effective conflict management mechanisms. Thus, it is also important to look at specific cases.

---

66. For further discussions on these environmental regimes, see Hayes and Zarsky, "Environmental Issues and Regimes in Northeast Asia," 298–306.

Chapter Fourteen

Symposium Summary

George Lindsey*

The Increasing Exploitation of Space

The commercial applications of space are increasing at a rapid pace, and there will be significant resistance to any measures of arms control which impose restrictions to its free exploitation. On the other hand, commercial interests will support steps to protect space assets from interference or destruction in times of conflict. Military activities are becoming dependent on services provided by civilian suppliers. An increasing number of applications are now for dual use. Repeated references have been made to the increasing dependence on the Global Positioning System (GPS), especially for control of both civilian and military air traffic and for the guidance of missiles. Before long, interruption of GPS would be catastrophic for air safety.

Space technology is proliferating, both horizontally and vertically. A field with spectacular advances is that of remote sensing. Electro-optical techniques are improving resolution toward the limits imposed by atmospheric turbulence and inhomogeneities. Synthetic Aperture Radar (SAR) is revealing unexpected information related to the roughness of reflecting surfaces, but progress is still needed in the extraction and interpretation of the data provided by the satellites.

In What Ways is Space Different from Land, Sea, and Air?

A dominant feature of space, without parallel in the other environments, is that satellites in LEO go around the world many times a day, passing over many countries every day.[1] Another is that the field of view from the high altitude of a satellite permits its sensors to observe, and its signals equipment to communicate, with locations on the ground, sea, or in the air — distributed over a substantial area of the Earth's surface.

---

* This chapter is an edited version of a presentation made to the Fifteenth Annual Ottawa NACD Verification Symposium, "Arms Control and the Rule of Law: A Framework for Peace and Security in Outer Space," 11–14 March 1998, Montebello, Quebec, Canada.

1. A satellite in a low orbit with an inclination of more than 50° will eventually cross above nearly every country in the world.

In the legal domain, a state holds complete sovereignty over its land, internal waters, territorial seas, and in the air above all of these, and a degree of economic sovereignty over the Exclusive Economic Zone (EEZ) of the oceans off its coast. But the high seas and space are international.

In the military domain, national armies demand complete control over the land of their country. National navies try to exert sea control and sea denial over their territorial waters,[2] often seek sea control over certain zones of the open ocean (such as sea-lanes-of-communication between allied countries), and sometimes sea denial. Somewhat less affected by geography, air forces seek to control the air above theatres of operations, perhaps extending into enemy airspace, and may manage to extend air superiority into complete air supremacy.

As far as aircraft are concerned, "air" is associated with the lower part of the atmosphere, which is able to provide buoyancy for aerostats, aerodynamic lift to keep aircraft aloft, and oxygen for combustion engines to propel them. Satellites in orbit proceed without any of these attributes, and are in space rather than in the air. Surface-to-surface ballistic missiles have to overcome air resistance during their boost phase, pass through space in ballistic flight in mid-course, and are slowed and heated by air drag as they reenter the lower atmosphere. The line of demarcation between air and space is not clearly defined.

In the case of space, one may ask what is implied by "space control." Does it imply the capability to operate space vehicles in spite of hostile efforts to prevent this (with the analogy of sea control)? Or does it imply the capability to prevent opponents from using space (like sea denial)? It may be that, unlike the maritime case, space denial is easier than space control. It may be easier to destroy the opponents' satellites than to protect one's own.

Because of the globe-circling character of satellite orbits, it would seem unlikely that national territories or geographical boundaries could be assigned. It might, however, be possible to distinguish layers of space such as those occupied by LEO and GEO (geosynchronous orbit) satellites.[3]

Multilateral Concerns over Security in Space

Since all nations border on space, and most space vehicles fly over most countries, the problems of security are necessarily of a multilateral nature.

However, the building and launching of a space vehicle is an extremely expensive undertaking, and demands a high level of technological and industrial sophistication, well beyond the national capacities of most states. In fact, the different degrees of space capability can be put into five categories:

---

2. Sea control implies the capability to use the relevant area of the ocean even if disputed by an enemy. Sea denial implies the capability to prevent an enemy from entering the area.
3. Regions that have already been identified for scientific purposes include the thermosphere, the ionosphere, and the Van Allen belts.

- autonomous for all missions;
- autonomous for some missions;
- significant capabilities, but partially dependent on other states;
- some capabilities, but largely dependent on other states; and,
- no capabilities.

Many of the discussions regarding strategic deterrence, strategic defence, ICBMs, and control of space were relevant only to the two members of the first category. Preservation and development of the commercial services provided by space technology concern all of the states in the first four categories. These include both protection from the hostile acts of enemies wishing to destroy satellites and assurance that otherwise cooperative partners will not withdraw services on which a state has become dependent. Defence against theatre ballistic missiles (TBMs) concerns members of all five groups. Negotiations of multilateral treaties on arms control concern all five. Concerns over control of space separate the paramount member of the first category from all the others.

Those states which could develop an anti-satellite weapon (ASAT) capability are likely to be those already dependent on space, and therefore with the greatest incentive to be able to protect satellites. However, there could be exceptions in the case of a "rogue" state more interested in blackmail or in destroying the satellites of their enemies. Although ASAT can be conducted from space, it can also be conducted from the surface.

Two Approaches to the Protection of Space

It has been possible to provide legal protection in times of peace for activities on land, sea, and air under national sovereignty, together with some steps to eliminate piracy on the high seas, but they provide little protection in times of war.

Effective protection of activities in space is not likely to be provided by declarations of intentions, or promises not to do things which are physically possible with certain equipment that exists or could be adapted fairly quickly. Instead, it would be necessary to arrange that the equipment required to interfere with space activities does not exist, and that measures to construct and deploy such equipment will become known and can be stopped.

Most of the basic technical problems stem from the unfortunate fact that satellites are extremely vulnerable to several types of weapons, some of them located in space, but others on the surface of the Earth, or sea, or in the air. These weapons have other roles which their owners will be reluctant to forego.

Some American authorities have declared space control to be a national objective. In fact, given the vulnerability of satellites to ASAT weapons, a power possessing sophisticated surface-based aerospace weaponry could be able to destroy its opponents' satellites (perhaps in LEO), which would be a form of space denial. But this would not in itself prevent adversaries from using their own surface-based weapons to destroy some, if not all, American satellites. This would mean that the United States had failed to exert complete space control.

It may well be that an American capability to deny the use of space to its opponents will never be used, only used as a form of unilateral deterrence, or used only for a purpose which carries the approval of most of the world — perhaps formally endorsed by the United Nations or a major alliance such as NATO. Perhaps this is the best solution to the protection of space, but it would be resented by many countries as a symbol of hegemony and inequality, as well as a rejection of the rule of law.

Another solution is to seek multilateral agreements to control the existence of equipment capable of inactivating, disabling, or destroying satellites. Such an approach will encounter formidable obstacles, some of them technical, some organizational. The national and organizational problems stem from the five different levels of capability and dependence on space. Those in the lower categories will be interested in the protection of whatever services are provided for them by other states. Those in the middle categories will be concerned with both their own and their collaborators' protection, but may have reservations about restrictions placed on their own weapons (likely not intended for ASAT use but nevertheless able to be converted for that purpose). And the country in the top category, whose absence would render multilateral agreements of limited usefulness, may prefer the unilateral solution.

The Feasibility and Potential Significance of a Treaty to Prohibit Weapons in Space

Possible spaceborne weapons could be used against three types of targets: other space vehicles in orbit, ballistic missiles in flight, and surface targets (on the ground, on the sea, or in the air). The most likely spaceborne ASAT weapons appear to be lasers.[4] Another possibility is small rocket-propelled homing missiles. It is also possible for the crew of a personned satellite to board, inspect, and, if so inclined, to disable another satellite.

The most important contribution to ballistic missile defence (BMD) that is made from space is early warning of missile launches, but this involves sensors rather than weapons. Space-based anti-ballistic missile weapons are already prohibited for the states party to the ABM Treaty. The "Brilliant Pebbles" of the Strategic Defense Initiative (SDI) programme would not have complied. However, a very likely form of defence against both intercontinental and theatre ballistic missiles will be to attack them during their boost phase, when they are most vulnerable.[5] The most likely weapon will be a laser

4. Low-powered lasers have many uses as instruments and means of communication. A laser weapon capable of destroying a ballistic missile would need very high power, more than adequate to disable a satellite. A treaty to prohibit ASAT laser weapons would have to allow the use of the lower-powered instrumental types.

beam, projected either from an aircraft or a satellite. Space-based assistance to BMD interceptor weapons on the ground would be through cueing, tracking, and possibly fire control.

There seems little reason to attack targets on Earth directly from space. The way in which satellites can affect combat on the ground or at sea is by collection and transmission of information, and assistance to accurate navigation. The most effective way for a satellite to disrupt sophisticated military systems indiscriminately, both in space and over a very large area of Earth, would be by detonation of a large nuclear explosion at an altitude selected to deliver an electromagnetic pulse, able to destroy electronic equipment dependent on semiconductor chips. This would be an infringement of the Outer Space Treaty.

The removal of weapons from space would not ensure the safety of satellites, especially those in LEO. They would still face the threat of destruction from surface-based interceptor missiles, whether built for ASAT, or adapted by attaching ASAT warheads to BMD rockets, or space launch vehicles. And, ASAT lasers could be located on the ground, at sea, or on aircraft. A more ambitious treaty, prohibiting all forms of ASAT weapons, would require significant restrictions to be agreed for the surface-based weapons just listed, and certainly for the development of theatre missile defences (TMD).

Ballistic Missiles and Ballistic Missile Defence

Although our focus here has been on arms control in outer space, the interrelationship between protection of activities in space and the weapons which threaten satellites has inspired some divergence to the subjects of ballistic missiles, defence against ballistic missiles, and their significance for global stability.

The situation for ICBMs is very different from that of TBMs. The number of ICBMs is not increasing, and can be significantly reduced if the START treaties are ratified and extended. But TBMs are proliferating, in hands not all inspiring confidence for future peace and stability. The largest numbers are of fairly unsophisticated designs with maximum ranges of 500 kilometres or less, but more advanced models are beginning to proliferate as well.

Theatre ballistic missiles beget the desire for ATBM defence. What the reverse motivations may be is not clear. Will appearance of ATBM defences cause a potential aggressor to reverse or forego acquisition of TBMs? Or will it induce that aggressor to build enough TBMs to overwhelm the defences? It could be reasoned that BMD is destabilizing for arms races, but TBMD would be stabilizing in a crisis.

---

5. In its boost phase a ballistic missile offers a slow-moving and vulnerable target, including the large propulsion stages with unburned rocket fuel. As soon as the missile rises above the lower atmosphere the hot gases of the rocket exhaust advertise its presence to infrared sensors thousands of kilometres away.

It has been pointed out that the real threat of ballistic missiles is strongly related to the payloads which they deliver. Nuclear is the most to be feared, on all counts. Biological may become a very serious threat, having important psychological, military, and economic effects. Chemical is possible today, with its effects most likely psychological and economic. A CBW attack could provide the motive for nuclear retaliation.

Considerable debate revolved around the ABM Treaty. A bilateral agreement, designed by the two superpowers at the height of the Cold War for limitation of defence against strategic ballistic missiles, it is in need of amendments if it is to continue to be a viable instrument. But, if it had not been agreed and observed, the world would be a more dangerous place today.

Some of the things that need to be addressed are the demarcation between ABM and TMD; laser weapons; and spaceborne, airborne, and shipborne components of military systems that could have an application for BMD.

Defence against TBMs represents the latest stage in the development of surface-based defence against flying objects, which began with anti-aircraft and then anti-cruise missile defence using guns, then added guided missiles, then extended their performance to be able to intercept crude TBMs. Now the capabilities of SAMs are being improved to provide "lower tier" systems able to protect targets within limited areas against more advanced TBMs, to be followed by "upper tier" systems giving protection to large areas. This progress in capabilities already gives the new BMD systems a limited ASAT capability against satellites in the lower LEOs, and there is every reason to expect this to increase in the absence of some form of control on TMD.

With booster rockets and guidance systems much more capable than those needed for short-range ballistic missiles (SRBMs) or TMD, ABM systems and space launchers could be used to propel ASAT warheads into much higher satellite orbits.

It should be remembered that as far as long-range surface-to-surface weapons are concerned, the threats to a state depend very much on the distance of its perceived potential enemies from its borders. The ABM Treaty was confined to defence against "strategic ballistic missiles," which the US and USSR defined for their purposes as having a maximum range of over 5,500 kilometres — a distance related to the geographical separation of their borders. But a country sharing a common border with a potential enemy could consider a missile with a very short range as posing a strategic threat to its territory.

One stabilizing multilateral step that could be taken very soon, with equipment now in use, would be to provide warning to any interested state of the launching of rockets anywhere in the world.[6]

---

6. Launches of rockets from known space launch and missile test ranges could be identified as probably not threatening, especially if previous announcements were made.

Partnerships and Processes to Support a Framework for Peace and Security in Outer Space

Lessons in garnering effective widespread public support for an arms control initiative can be derived from the steps which led up to the recently completed treaty on antipersonnel landmines. Although protection of satellites would not provide the compelling cases of humanitarian tragedies so well portrayed by the images of landmine victims, the subject should have a direct appeal to many commercial interests in Canada and elsewhere. With the objective of a multilateral agreement, it would be best to seek the support of international, rather than national NGOs. A technical advisory group could be useful, provided that the political will was present to press for success.

While the Conference on Disarmament seems to be the obvious agency in which to discuss and negotiate possible agreements regarding arms control in space, the pace of progress in this body has been disappointingly slow. An attempt to amend the Outer Space Treaty was not thought to offer good prospects. The best hope seems to be to seek a mandate from the CD to discuss the prospects for a new multilateral agreement.

Another approach could be to create an ad hoc organization (like the MTCR), establish a clearly-defined objective, and muster wide support before seeking to negotiate a treaty. Rather than aiming for an early comprehensive treaty, it would be better to concentrate on a limited, finite step in the right direction. The prohibition of weapons in space would be such an objective. It would have the advantage that there are no weapons in space now to be dismantled. It would afford some protection to satellites, though it would have to be acknowledged right from the beginning that they would still be at risk from surface-based weapons. Progress is more likely to be achieved step-by-step than by attempting one giant leap.

Part III
Working Group Reports

Chapter Fifteen

Working Group One Rapporteur's Notes: Legal and Policy Concerns

Peter Gizewski*

Recent years have witnessed growing interest in the role of outer space in international security. This is not surprising. All nations border on space and are increasingly dependent upon the heavens for the conduct of a variety of military and commercial activities — i.e., communication, warning, and reconnaissance. Space-based activity has increased dramatically in the past two decades.

In general, the vast majority of activity is benign, and the existing legal regime governing outer space has gone largely unchallenged. Yet, as capabilities expand and interests in this "final frontier" grow, activities may eventually emerge which threaten both national and international security. In time, technological diffusion may promote a degree of competition and conflict sufficient to worry even the most sophisticated space-faring nation. The range of possible activities is wide — i.e., ground-to-space, space-to-ground, and space-to-space. And the potential threats posited are considerable.

Such possibilities raise a variety of questions concerning the present legal regime. Specifically: of what does the regime consist? Is it adequate for addressing present and future threats? If not, does it require revision or merely amendment? Finally, what measures are necessary to better ensure that it provides effective security in the future? The following discussion addresses these questions, and offers some general suggestions for ensuring a sound regime in the future.

The Current Legal Context

Today, outer space is the subject of a range of multilateral and bilateral agreements, rules, and norms dealing both directly and indirectly with its regulation and the activities which occur within it. The current regime includes:

---

* This chapter is an edited version of rapporteur notes recorded and compiled by Peter Gizewski at the Fifteenth Annual Ottawa NACD Verification Symposium, "Arms Control and the Rule of Law: A Framework for Peace and Security in Outer Space," 11–14 March 1998, Montebello, Quebec, Canada.

- the Outer Space Treaty;
- the Liability Convention;
- the Rescue Agreement;
- the Registration of Space Vehicles;
- the Moon Agreement;
- the ABM Treaty; and,
- the Agreement on Non-Interference.

The current rules include: prohibitions against placing in orbit and testing both nuclear weapons and anti-ballistic missiles, deploying military bases on celestial bodies and orbits, and using outer space for the purpose of modifying the environment for hostile purposes. Beyond this, the US-Russia noninterference agreement obliges both parties to refrain from interfering (for military purposes) with one another's space-based remote sensing activities.

Gaps in Coverage

Unfortunately, the present regime is characterized by loopholes, imprecision, and limitations *vis-à-vis* the military activities that are actually covered. For instance, while the Outer Space Treaty offers some assurances against the weaponization of space (Article IV), its terms fail to address conventional and certain varieties of exotic weaponry (e.g., certain anti-satellite (ASAT) technologies), the deployment of weapons of mass destruction in partial — as opposed to full — orbits, and vehicles capable of functioning between airspace and outer space. Meanwhile, other agreements, such as the ABM Treaty, are bilateral and thus exceedingly limited in their coverage.

Addressing such shortcomings and the dangers they pose is a formidable task. Certainly, states may choose to defend space-based assets through the deployment of military power. Yet, while this may offer an adequate short-term solution, its long-term effectiveness is less certain. As technologies spread, potential adversaries may multiply, increasing threats posed as well as the costs associated with deterring them. The result may well be less security at greater economic cost.

In fact, an effective long-term policy demands efforts aimed at strengthening the legal regime. By bolstering the framework for international cooperation, states may achieve greater long-term security at lower cost, avoiding the competitive arms buildups which could ultimately characterize excessive reliance on military power.

Strengthening the Regime

An ideal regime would be universal in scope, verifiable, and capable of assuring the community of states that important national assets will be protected and transgressions addressed. In the case of outer space, this is particularly demanding. Existing rules and norms are too narrow and incomplete to inspire state confidence in their capacity to guard against future threats. Furthermore, filling the gaps in the various agreements which presently exist promises to be an arduous process.

For instance, while coverage of potential threats in the Outer Space Treaty could be improved by revising the non-weaponization clause to include "all" weapons in space, this would create a two-tiered system — a change not likely to appeal to many states. Other agreements, such as the ABM Treaty, would require a process of multilateralization which is, at present, impractical.

A new regime, carefully devised to avoid the shortcomings of existing arrangements may provide a preferable alternative. The regime would draw upon arms control, disarmament, and confidence building institutions and measures in other areas in an effort to devise the most effective "new" arrangement possible.

Construction of such a regime would require intensive investigation. Based on present realities however, a number of options are worthy of consideration. Policy-makers might consider devising a "Stockholm for Space" modelled on the principles and modalities of the existing Stockholm Agreement. They might encourage the development of a joint declaration of permitted and prohibited activities by key interested parties, along the lines of the declaration on terrorism reached in Bonn. Beyond this, efforts could focus on establishing a formal regime composed of specific restraints. Elements of such a regime would include:

- a broadening of the Outer Space Treaty prohibition (Article IV) to enshrine the general "non-weaponization" of space;
- a noninterference clause which would prohibit parties from obstructing one another's peaceful assets and activities in space (along the lines of that governing the US and Russia);
- a commitment not to launch weapons in space for other parties;
- a pledge not to test weapons in space (although research and development would be permitted); and,
- the development of penalties for noncompliance.

Provisions for verifying compliance with the regime would also require elaboration, and would involve: state declarations of openness, the gradual development of on-site inspection arrangements, an expansion of space surveillance (perhaps leading to a multilateral framework), and the development of generic transparency arrangements. A standing consultative commission charged with investigating allegations of noncompliance and updating the regime in light of evolving technology could also be created.

Other issues would demand attention as well. Guarding against the emergence of a two-tiered system would continue to represent a concern. Defining the regime's main purpose — i.e., whether it should be directed principally at non-weaponization or at the peaceful uses of space — would be essential. And questions concerning how such a regime should treat the development and management of new military technologies — i.e., theatre missile defences, electronic warfare — will linger. Yet, such issues would arise even in the absence of a renewed effort at regime building, and the capacity to effectively address them under the present framework is not self-evident.

Conclusion

Movement toward a new legal regime will require support and participation from the international community. Particularly significant will be the attitude of the United States — currently the premier space-faring nation in the international system. Given the growing importance of space to US national security, temptations to de-emphasize pursuit of an improved legal regime in favour of a unilateral approach based on military power may be high, particularly in the near term.

Yet, over the long run, the exercise of restraint may prove to be the wiser course. Indeed, it may be preferable to co-opt emerging and potential future space powers into a system of cooperation now than face the atmosphere of competition, suspicion, and the rising defence costs which excessive reliance on a policy of deterrence based on military power may breed later. Convincing policy-makers of this fact may well prove to be the most crucial challenge facing efforts to ensure a more effective regime in the years to come.

Chapter Sixteen

Working Group Two Rapporteur's Notes: Arms Control in Outer Space — The Multilateral Dimension

Jean-Philippe Tachdjian*

Introduction

The United States and Russia, the world's two great space powers, have traditionally kept the issue of space and space-based weapons as their exclusive jurisdiction. They are continuing, albeit at a very slow pace, their bilateral talks on the matter of arms control in outer space.

The United States, however, is not prepared as yet to commit itself to the multilateral option for eliminating the use of weapons in outer space. For its part, Russia is keeping its options open on both bilateral and multilateral processes.

With the end of the Cold War, it seems that multilateralism has a greater opportunity for success than ever before. For the first time, a multilateral treaty banning the weaponization of outer space is a real, rather than remote possibility.

Useful Fora

During its discussions, the Working Group identified five possible fora for the development of such a treaty:

- the United Nations General Assembly (UNGA);
- the United Nations Disarmament Commission (UNDC);
- the United Nations Committee on the Peaceful Uses of Outer Space (COPUOS);
- the Conference on Disarmament (CD); and,

---

* This chapter is an edited version of rapporteur notes recorded and compiled by Jean-Philippe Tachdjian at the Fifteenth Annual Ottawa NACD Verification Symposium, "Arms Control and the Rule of Law: A Framework for Peace and Security in Outer Space," 11–14 March 1998, Montebello, Quebec, Canada. The opinions expressed in this chapter are those of the author and do not necessarily reflect those of the Department of Foreign Affairs and International Trade or the Government of Canada.

- an Ottawa-like alternative process to a normal diplomatic forum.

The members of the Working Group agreed that the Conference on Disarmament is the most likely place for such a treaty to develop. However, all other venues, including the UNGA and COPUOS, must be used to discuss and promote the issue.

It was also agreed that an alternative process, similar in type to that used to secure the Ottawa treaty banning land mines, should be explored only if and when other venues are exhausted, due to the normal difficulties of getting all relevant states, particularly the United States and Russia, on board.

It was proposed that the Conference on Disarmament name a Special Coordinator on the issue and create an ad hoc committee with a mandate to negotiate a treaty.

If negotiations do not form part of the mandate due to the opposition of the United States or another state, a mandate to discuss the issue should be sought. If a mandate to discuss the issue is impossible to secure due to the opposition of any state, Canada should suggest that a group of scientific experts be established to study the various aspects of verification of weapons in outer space.

Proposed Timetable for the Plan of Action

Without establishing target dates that may limit the ability to negotiate the issue, the proposed timetable would be as follows:

First stage:

- ad hoc committee created;
- preliminary papers are prepared; and,
- a draft agreement is prepared and circulated within the committee.

Second stage:

- the issue is discussed in the CD ad hoc committee, with the draft agreement used as the basis for discussion/negotiations.

Third stage:

- come to an agreement and sign a treaty on the matter; or,
- conclude that such an agreement is not possible in the CD in the near future and move toward an alternative venue, recognizing the difficulties identified earlier.

The "Quick Fix:" Amending the 1967 Outer Space Treaty

The Working Group also discussed the possibility of re-opening the 1967 Outer Space Treaty. However, after a discussion of Dr. Lucy Stojak's paper, a study of the Treaty's amending procedure and a review of the history of COPUOS, it was concluded that such a "quick fix" solution would not be feasible.

However, the members of the Working Group also agreed that all COPUOS venues should be used to promote and support the establishment of such a treaty and, in particular, the 1999 United Nations International Space Conference (UNISPACE III).

The Scope of the Proposed Treaty

The Working Group discussed at length the issue of the precise scope of the proposed treaty, in particular whether it should ban "weapons in space" (narrow scope — limited to weapons actually located in space) or "weapons for use in space" (broader scope — including ground-, sea-, or air-based weapons that may reach or pass through space on their way to their intended targets).

The Working Group concluded that the desirable objective would be an instrument based on the broadest scope possible. However, it was also agreed that, in practical terms, a treaty reflecting the narrower scope would be more likely to gather the needed support. As a result, the Group would begin by proposing a narrower instrument on the non-weaponization of outer space only. The broader scope may be added later, either during the negotiations or as a protocol to the narrower agreement.

The Verification Regime

The Working Group briefly discussed the issue of an adequate verification regime and identified three possible methods:

- verification in space, using space assets;
- verification on the launch pad, using on-site inspectors; and,
- verification on the assembly line, using on-site inspectors.

These three methods are not mutually exclusive and could be used in conjunction with one another.

The question of who would carry out the verification and create a data base of the inspection results was also briefly addressed. The use of an existing appropriate verification entity was preferred to the creation of a new entity.

Conclusions

The Working Group concluded that, for this process to work, the support of a broad range of like-minded but regionally diverse countries would be needed, as would the support of non-state actors and the public at large.

Chapter Seventeen

Working Group Three Rapporteur's Notes: Outer Space and Strategic Stability — The ABM Treaty and Ballistic Missile Defence

Steve James*

Question 1: What are the implications of current ballistic missile defence developments respecting strategic stability?

Observations

- Nuclear disarmament is the Conference on Disarmament's top priority — the concern is twofold: legacy superpower arsenals and the proliferation of strategic weapons.
- Strategic stability involves traditional superpower relationships, as well as regional considerations where homelands are potentially threatened by regional ballistic missiles.
- Participants differed as to whether US BMD initiatives, if ABM Treaty compliant, are stabilizing or de-stabilizing in nature. However, most agreed non-Treaty compliant development would be destabilizing.
- One observer noted that all BMD concepts inherently have some destabilizing characteristics; that is, BMD deployment inherently disrupts existing strategic balances and, thereby, may provoke renewed arms buildups.
- It was observed that concerned states may wish to propose dialogue on the subject of regional BMD deployments.

Conclusion

Widespread views exist as to the potential strategic stability implications of current BMD developments.

---

* This chapter is an edited version of rapporteur notes recorded and compiled by Steve James at the Fifteenth Annual Ottawa NACD Verification Symposium, "Arms Control and the Rule of Law: A Framework for Peace and Security in Outer Space," 11–14 March 1998, Montebello, Quebec, Canada. The opinions expressed in this chapter are those of the author and do not necessarily reflect those of the Department of National Defence or the Government of Canada.

Question 2: What are the implications of current ballistic missile defence developments and deployments respecting arms control in outer space?

Observations

- Space is currently militarized, but not weaponized.
- Only space-based WMD are currently prohibited by the 1967 Outer Space Treaty.
- Future BMD developments and deployments could further militarize outer space.
- Expanding commercialization of space heightens the domain's international importance.
- The deployment of space-based weapons is not necessary for current limited BMD concepts, but the long-term pursuit of an effective defence may encourage space weaponization in the future.
- Even Treaty-compliant BMD deployment could initiate an arms race in space; however, some observers noted, there are significant economic and technical near-term barriers that mitigate against that possibility.

Conclusion

Proponents of space-based BMD weapons will naturally oppose any efforts to ban weapons in outer space.

Question 3: Is there a role for multilateral participation in ballistic missile defence?

Observations

- The sharing of early warning data and notice of launches already exists.
- International participation in missile warning and space tracking systems has been proposed.
- Such initiatives have strong confidence and security building characteristics.
- There are already multilateral initiatives in BMD.

Conclusion

There does appear to be a role for multilateral, or even global, participation in early warning; that participation does not have to be related to a specific BMD initiative.

Recommendation

Multilateral sharing of early warning data and notice of launches should be further encouraged.

Question 4: Is there a role for multilateral limitations on ballistic missile defence?

Observations

- The *current* ABM Treaty is bilateral in nature and open to clarifications, modifications, and amendments.
- There is a role for limits on certain aspects of BMD — i.e., space-based weapons.
- BMD concerns and initiatives are taking place outside the traditional superpower relationship.
- Akin to the stability question between the superpowers, excessive BMD at a regional level *could* create a localized arms buildup.

Conclusion

There is a role for multilateral limitations on BMD.

Recommendation

Further study should be considered.

Question 5: Is there a linkage between the issues of ballistic missile defence and ASAT?

Observations

- High-velocity BMD systems inherently have ASAT capabilities.
- ASATs are not limited by any treaty.
- For verification purposes, discriminating between BMD and ASAT roles is problematic.
- Some laser technologies proposed would threaten LEO and GEO satellites.
- It may be difficult to verify the intended purpose of a deployed laser system.

Conclusion

A prohibition on the testing and deployment of ASATs, supported by a verification regime, would be desirable, as it would limit the military confidence in the effectiveness of any BMD system used in an ASAT role. An ASAT test ban could be a useful arms control measure.

Recommendation

Further study should be considered.

Chapter Eighteen

Working Group Four Rapporteur's Notes: Building a Constituency to Inform the Debate

Peter Stibrany*

Recommendations

i. Rearticulate and publicize Canada's essential interest in a more comprehensive civil space regime — why it is important economically, militarily, and morally.

ii. Identify key allies within the Conference on Disarmament (CD). Some common interest groups may be:

- international law associations;
- commercial interests;
- strategic analysis organizations; and,
- scientific societies.

iii. Investigate lessons from past situations:

- Law of the Sea;
- Comprehensive Test Ban Treaty (CTBT) Group of Scientific Experts (GSE) experience; and,
- ITU example for coordination of satellite frequencies.

iv. Create staged approach. Perhaps one milestone could be the millennium General Assembly, at which such a forward-thinking initiative would be welcomed.

v. Publish past studies on the Internet or CD ROM to researchers and interested parties.

Questioning the Assumptions

i. Does the Canadian Government have the will to carry through a campaign in this area? There are some indications that it would:

---

* This chapter is an edited version of rapporteur notes recorded and compiled by Peter Stibrany at the Fifteenth Annual Ottawa NACD Verification Symposium, "Arms Control and the Rule of Law: A Framework for Peace and Security in Outer Space," 11–14 March 1998, Montebello, Quebec, Canada.

- it *is* the current Canadian policy; and,
- it is very much in Canada's long-term interest.

Such a campaign would probably need to build support from interests outside the Canadian government.

ii. Is non-weaponization of space good? Yes, but the issue should be restated as a positive goal — for example, the "protection of space assets" or "the protection of space for peaceful purposes."

iii. Is building a bigger constituency good? Yes, but this is not a populist issue, and will not likely become one for a long time. Constituency building will need to be done in more specific, targeted areas.

iv. Is there growing acceptance of multilateral arms control and disarmament (ACD)? Yes and no:

- multilateral agreements have become more prevalent;
- the traditional "duopoly" issues of nuclear weapons and ballistic missile defence (BMD) remain so; and,
- a supplier cartel, such as the Missile Technology Control Regime (MTCR), is more acceptable to the US than multilateral arms control with respect to ballistic missiles and probably with respect to the space area.

Roles and Interests

What are the roles of government, the military, and non-governmental groups (including advocacy groups, academia, the media and the business sector) regarding arms control processes, and have these roles changed?

a. Since the end of the Cold War, commercial demands have more influence:

- high-resolution imagery;
- export of satellites; and,
- use of foreign launch vehicles.

These more powerful commercial interests are not really focussed as yet on arms control issues. On balance, they will likely be in favour of ACD. That said, government policy in the United States is still driven largely by military interests.

b. Non-governmental organizations (or more generically, non-state actors) have a larger role because:

- government retrenchment;
- they can influence the societies of other countries directly and thus pressure other governments;
- liberalization (as in China); and,
- more diversity of interests involved.

New Instruments and Partnerships

How can new instruments and partnerships be applied and promoted in terms of arms control in outer space and what considerations need to be borne in mind (i.e., advantages, disadvantages, opportunities, limitations)?

a. Infobuilding — government creating a more informed community of interest.

b. Group of Scientific Experts — not new in general, but new for space issues.

c. Partnership with Industry:
   - ensure stable commercial environment; and,
   - "free trade" regime instead of "protected trade" regime.

d. Limitations:
   - someone needs to take organizing/motivating role;
   - issue will not run itself;
   - government needs to foster creation of clear vision of objective;
   - no tokenism or co-opting; and,
   - available resources.

e. It is uncertain as to whether part of ACD monitoring should be tied to a general civil organization for:
   - space traffic monitoring; and,
   - debris management.

   The character of the relationship with COPUOS (Committee on the Peaceful Uses of Outer Space) on this issue also remains to be determined.

Multilateral Solutions

Multilateral ACD does work (as demonstrated in other areas) and ultimately can offer the best long-term solutions to space security as well. From the point of view of the United States, the overwhelming space power today, this may be the best time to negotiate multilateral space ACD from a position of strength — leverage will only decline with the passage of time.

List of Abbreviations and Acronyms

ABM	anti-ballistic missile
ACDA	United States Arms Control and Disarmament Agency
AICBM	anti-intercontinental ballistic missiles
ALPS	accidental launch protection system
ASAT	anti-satellite (weapons)
ATBM	anti-tactical ballistic missile
BMD	ballistic missile defence
BMDO	Ballistic Missile Defense Organization
BTWC	Bacteriological (Biological) and Toxin Weapons Convention
CBM	confidence-building measure(s)
CCW	Convention on Certain Conventional Weapons
CD	Conference on Disarmament
CFE	Treaty on Conventional Armed Forces in Europe
CMC	Cooperative Monitoring Center
COPUOS	UN Committee on Peaceful Uses of Outer Space
CSBM	confidence and security building measure(s)
CTBT	Comprehensive Test Ban Treaty
CWC	Chemical Weapons Convention
DoD	Department of Defense (US)
DoT	Department of Transportation (US)
DPRK	Democratic People's Republic of Korea
Fizz	fissionable material "cutoff" treaty
GEO	geo-synchronous orbit
GPS	Global Positioning System
GSE	Group of Scientific Experts (CD)
HEU	highly-enriched uranium
HUMINT	human intelligence
IAEA	International Atomic Energy Agency
ICBM	intercontinental ballistic missile
IGA	intergovernmental agreement
INF	Intermediate-Range Nuclear Forces Treaty
ISMA	International Satellite Monitoring Agency
LEO	low Earth orbit
LPAR	large phased array radar

LTBT	Limited Test Ban Treaty
MAD	mutual assured destruction
MARV	manoeuvrable reentry vehicle
MIRACL	mid-infrared advanced chemical laser
MIRV	multiple independently targeted reentry vehicle
MOU	memorandum of understanding
MTCR	Missile Technology Control Regime
NACD	non-proliferation, arms control, and disarmament
NAM	Non-Aligned Movement
NATO	North Atlantic Treaty Organization
NBC	nuclear, biological, chemical
NGO	non-governmental organization
NNWS	non-nuclear weapons state
NORAD	North American Aerospace Defence Command
NPT	Nuclear Non-Proliferation Treaty
NST	nuclear and space talks
NTM	national technical means
NTSC	non-traditional security challenges
NWS	nuclear weapons states
NWFZ	nuclear-weapons-free zone
OMV	ongoing monitoring and verification
OPCW	Organization for the Prohibition of Chemical Weapons
OSCE	Organization for Security and Cooperation in Europe
OSI	on-site inspection(s)
P-5	Group of 5 Permanent Members of the UN Security Council
REVCON	Review Conference
SALT	Strategic Arms Limitation Treaty or Talks
SAM	surface-to-air missile
SCC	Standing Consultative Commission
SDI	Strategic Defense Initiative
SDIO	Strategic Defense Initiative Organization
SLBM	submarine-launched ballistic missile
SRBM	short-range ballistic missile
START	Strategic Arms Reduction Treaty
TBMBT	total ballistic missile ban treaty
TMD	theatre missile defence

UNDC	United Nations Disarmament Commission
UNGA	United Nations General Assembly
UNISPACE	United Nations Conference on the Exploration and Use of Outer Space
UNSCOM	United Nations Special Commission on Iraq
UNSSOD	United Nations Special Session on Disarmament
VHR	very high resolution remote sensing system
WMD	weapons of mass destruction

Ambassador Sidney N. Graybeal, Dr. Patricia Bliss McFate, and F.R. (Ron) Cleminson at Fifteenth Annual NACD Verification Symposium.

Ambassador Sidney Graybeal, In Memoriam

For nearly a decade, Ambassador Sidney Graybeal was an active participant in each of the Annual Ottawa Verification Symposia hosted by York University as part of the Verification Research Program of the Department of Foreign Affairs and International Trade. As one of the most knowledgeable and capable "arms controllers" in Washington, Ambassador Graybeal had an extraordinary record of service to his country. Beginning as a B-29 pilot in the Pacific theatre of operations during the Second World War, Sidney continued in government service, first with the Central Intelligence Agency and later in a number of arms control capacities, most notably with respect to the SALT process and the ABM Treaty. He was a member of the US delegation to the 1958 Surprise Attack Talks and served as the first American Commissioner on the Standing Consultative Commission, established to address questions concerning ABM Treaty implementation. Honoured by several American Presidents, he remained active with Science Applications International Corporation (SAIC) after his career in government service.

His association with Canadians in the areas of international security and arms control was of great value and his personal friendship will live on with those of us who knew him. Ambassador Graybeal and his wife, Dr. Patricia McFate, along with a number of distinguished Canadian researchers and scholars, co-authored a series of four major studies relating to multilateral verification and the non-proliferation, arms control, and disarmament (NACD) process.

Included in this proceedings is a paper submitted and presented by Ambassador Graybeal and Dr. McFate. In addition, Sidney chaired a series of presentations during the symposium and participated throughout the sessions, providing customarily incisive and evocative comments. During the concluding session, he enthralled the audience with an impromptu glimpse of his vision of the NACD world ahead.

It was, therefore, with profound sadness that we learned of Ambassador Graybeal's death at home in Santa Fe less than a week after his participation in the symposium. We cherish our wonderful memories of Sidney among his friends and colleagues. Sidney will be remembered variously as an officer and a gentleman, a distinguished patriot, and an accomplished and skilled negotiator. We would prefer to remember him simply as a very valued friend. Our hearts have been and will be with his family, and particularly with his wife, Patricia McFate, also a very valued friend and colleague.

per ardua ad astra

Other Publications by the Centre for International and Security Studies

The following is a listing of monographs published by the Centre for International and Security Studies dealing with non-proliferation, arms control, disarmament, and verification issues. For information on how to obtain these publications, please contact: Centre for International and Security Studies, York University, 4700 Keele Street, Toronto, Ontario, Canada M3J 1P3, Tel: (416) 736–5156, Fax: (416) 736–5752, E-mail: yciss@yorku.ca, Internet: http://www.yorku.ca/research/ciss

Non-Proliferation, Arms Control and Disarmament: Enhancing Existing Regimes and Exploring New Dimensions. Proceedings of the 1997 Canadian Non-Proliferation Workshop, Peter Gizewski, ed., 1998.

Cyberspace and Outer Space: Transitional Challenges for Multilateral Verification in the 21st Century. Proceedings of the Fourteenth Annual Ottawa NACD Verification Symposium, J. Marshall Beier and Steven Mataija, eds., 1997.

Non-Proliferation Agreements, Arrangements and Responses. Proceedings of the 1996 Canadian Non-Proliferation Workshop, Andrew Latham, ed., 1997.

Verification, Compliance and Confidence-Building: The Global and Regional Interface. Proceedings of the Thirteenth Annual Ottawa NACD Verification Symposium, J. Marshall Beier and Steven Mataija, eds., 1996.

Multilateral Approaches to Non-Proliferation. Proceedings of the 4th Canadian Non- Proliferation Workshop, Andrew Latham, ed., 1996.

Proliferation in All Its Aspects Post-1995: The Verification Challenge and Response, J. Marshall Beier and Steven Mataija, eds., 1995.

Moving Beyond Supplier Controls in a Mature Technology Environment. Proceedings of the Third Canadian Non-Proliferation Workshop, David Mutimer, ed., 1995.

Non-Proliferation and Multilateral Verification: The Comprehensive Nuclear Test Ban Treaty (CTBT), Steven Mataija, ed., 1994.

Reconsidering Confidence and Security Building Measures: A Critical Analysis, Andrew Richter, 1994.

Control But Verify: Verification and the New Non-Proliferation Agenda, David Mutimer, ed., 1994.

Proliferation and International Security: Converging Roles of Verification, Confidence Building and Peacekeeping, Steven Mataija and Lyne C. Bourque, eds., 1993.

Multilateral Verification and the Post-Gulf Environment: Learning From the UNSCOM Experience, Steven Mataija and J. Marshall Beier, eds., 1992.

Towards Helsinki 1992: Arms Control in Europe and the Verification Process, Heather Chestnutt and Steven Mataija, eds., 1991.

International Verification Organizations, Ellis Morris, ed., 1991.

Verifying Conventional Force Reductions in Europe: CFE I and Beyond, Heather Chestnutt and Michael Slack, eds., 1990.